# BASIC EQUATIONS

## BY S.H. COLLINS

COVER DESIGN BY KATHY KIFER

Emily and Elena

Published by
**Garlic Press**
605 Powers St.
Eugene, OR  97402

ISBN 0-931993-84-9
Order Number GP-084

www.garlicpess.com

# Table of Contents

# Chapter 1:
# Equations & Number Relationships

An **equation** is a mathematical sentence containing an equal sign.  The value of the numbers on the right side of the equation must equal the value of the numbers on the left side of the equation, if the equation is to be true.

$$6 + 3 = 9 \qquad\qquad 27 \times 3 = 81$$

$$6 = 42 \div 7 \qquad\qquad 100 = 150 - 50$$

## Part 1 — Writing Complete Equations

Regular computational math problems can be written as complete equations.

### ◆Examples

**Math Problem ⟹ to ⟹ Equation**

$$\begin{array}{r} 152 \\ + 17 \\ \hline 169 \end{array}$$
$$152 + 17 = 169$$

$$24\overline{)216}\,^{9}$$
$$216 \div 24 = 9$$

$\frac{1}{2}$ of 6 is 3
$$\frac{1}{2} \times 6 = 3$$

$$\begin{array}{r} \$17.95 \\ - \ 12.56 \\ \hline \$5.39 \end{array}$$
$$\$17.95 - \$12.56 = \$5.39$$

Once a simple math problem is solved, an equation can be written to explain the solution:

### ◆Examples

**Problem ⟸ Solution ⟹ Equation**

$$\begin{array}{r} 64 \\ \times \ 9 \\ \hline \end{array} \qquad \begin{array}{r} 64 \\ \times \ 9 \\ \hline 576 \end{array} \qquad 64 \times 9 = 576$$

## Problem

Elena sharpened the 24 pencils in her new set. By the end of the day, 17 remained sharp. How many must be resharpened?

**Solution ➤ Equation**

$$\begin{array}{r} 24 \\ -\ 17 \\ \hline 7 \end{array}$$

$$24 - 17 = 7$$

**Problem ➤ Solution ➤ Equation**

$$35\overline{)26.25}$$

$$35\overline{)26.25}^{\ .75}$$

$$26.25 \div 35 = .75$$

---

## Exercise 1 · Write an equation for these math problems.

**1.** $\begin{array}{r} \frac{2}{3} \\ -\ \frac{4}{9} \\ \hline \frac{2}{9} \end{array}$

$\frac{2}{3} - \frac{4}{9} = \frac{2}{9}$

**2.** 42% of 93 is 39.06

$.42\overline{)93.0}^{\ 39.06}$

**3.** $\frac{3}{4}$ equals $\frac{9}{12}$

$\frac{3}{4} = \frac{9}{12}$

**4.** If Al drives for 8 hours and travels 720 km, his average rate per hour is equal to 90 km. (Hint: What balances 90?)

$6\overline{)720\,km}^{\ 90\,km}$

**5.** $\frac{1}{2}$ and $\frac{1}{4}$ equal .75

$\frac{1}{2} + \frac{1}{4} = \frac{3}{4}$

**6.** $\begin{array}{r} 97\frac{5}{12} \\ -\ 8\frac{7}{8} \\ \hline 88\frac{13}{24} \end{array}$

**7.** 6 is 50% of 12

**8.** A \$29.95 appliance is discounted \$5.99. Its sale price is \$23.96.

**9.** $\begin{array}{r} 127 \\ \times\ 52 \\ \hline 6604 \end{array}$

**10.** $\begin{array}{r} 132 \\ 19 \\ +\ 7 \\ \hline 158 \end{array}$

**11.** $92\overline{)15{,}640}^{\ 170}$

12. The sides of a square are 2 m.  The distance around the square is 8 m.

13. Ace worked 3 hours 20 minutes on Monday and 4 hours 30 minutes on Tuesday.  He worked 7 hours 50 minutes altogether.

14. Mrs. March's monthly salary is $2400.  Working for 12 months, her yearly salary is $28,800.

## Part 2 — Writing Equations after Finding an Unknown

In order to be balanced and true, an equation must have a solution.  The solution to an equation is **unknown** until the problem is solved.  When the problem is solved, all parts of the equation are present.

### ◆xamples

- Two friends went to lunch.  One spent $4.95 and the other spent $5.25.  **How much did they spend altogether?**

| Statement of Problem | Solution of Problem | Writing a Balanced Equation |
|---|---|---|
| $4.95 <br> + $5.25 <br> ——— <br> = Amount spent altogether (unknown) | $4.95 <br> + $5.25 <br> ——— <br> $10.20 | $4.95 + $5.25 = $10.20 |

- What is 15% of 17.84?

| Statement of Problem | Solution of Problem | Writing a Balanced Equation |
|---|---|---|
| 17.84 <br> × .15 <br> ——— <br> = 15% of 17.84 (unknown) | 17.84 <br> × .15 <br> ——— <br> 2.6760 | .15 × 17.84 = 2.676 |

**Exercise 2** Solve these problems and write complete equations.

1. 502.62
   27.10
   + 3.86

2. $5.2\overline{)14.56}$

3. $\frac{3}{4}$
   $-\frac{1}{6}$

4. How much is 1520 take away 1162?

5. What is $\frac{1}{5}$ of 75?

6. 7 is what percent of 10?

7. A designer shirt that regularly sells for $39.95 is on sale for $24.29. What is the savings?

8. A loan of $3000 has a yearly interest rate of 15%. What is the interest for one year?

9. Emily ran $\frac{2}{3}$ of the distance to the top of Spencer's Butte. The distance to the top is $1\frac{1}{2}$ miles. How far did Emily run?

10. Together three friends earned $57. If they divided the money equally, how much did each receive?

11. What is 127 multiplied by 12?

12. How many times will $1\frac{5}{8}$ go into $\frac{7}{16}$?

13. Nine hours equals how many minutes?

14. There are 97 registered voters on our block. Only 42 voted. How many did not vote?

15. $\dfrac{\$9.50 + \$7.30}{2} =$

# Number Properties and Operations

Certain properties and operations influence how equations are solved. Whether working with numbers or variables (which you will encounter in Chapter 2), knowledge and mastery of these certain properties and operations is indispensable.

## A. Commutative Property—Changing Order

• The order that numbers are added does not change their sum.

• The order that numbers are multiplied does not change their product.

| Numbers |
| --- |
| EXAMPLES |

| Whole Numbers | Integers |
| --- | --- |
| 1, 2, 3, 4... | −1, −2, −3, −4... |
| Rational Numbers | Irrational Numbers |
| 5.6, $\frac{1}{3}$, $-\frac{1}{5}$ | $\sqrt{2}$, $-\sqrt{7}$, $\pi$ |

### ◆xamples

| | | |
| --- | --- | --- |
| $12 + 6 = 18$ | is the same as... | $6 + 12 = 18$ |
| $\frac{1}{3} + \frac{1}{4} = \frac{7}{12}$ | is the same as... | $\frac{1}{4} + \frac{1}{3} = \frac{7}{12}$ |
| $3.75 \times 9.1 = 34.125$ | is the same as... | $9.1 \times 3.75 = 34.125$ |
| $5\frac{1}{2} \times 2\frac{1}{4} = 12\frac{3}{8}$ | is the same as... | $2\frac{1}{4} \times 5\frac{1}{2} = 12\frac{3}{8}$ |

Note that the Commutative Property does not hold for subtraction or division.

**Simple Proof** of the Commutative Property, using a balanced equation:

$$15 + 6 = 6 + 15 \qquad\qquad \$2.75 \times 6 = 6 \times \$2.75$$
$$21 = 21 \qquad\qquad\qquad \$16.50 = \$16.50$$

Why the Commutative Property doesn't work for subtraction or division:

$$15 - 6 \neq 6 - 15 \qquad\qquad 32 \div 4 \neq 4 \div 32$$
$$9 \neq -9 \qquad\qquad\qquad 8 \neq \frac{1}{8}$$

**Exercise 3** Use the Commutative Property to prove the equality (or inequality) in these problems.

**1.** $75 + 9 = 9 + 75$

**2.** $5 \times 6 = 6 \times 5$

**3.** $\frac{5}{8} \div \frac{3}{4} = \frac{3}{4} \div \frac{5}{8}$

**4.** $.5 \times .96 = .96 \times .5$

**5.** $47.962 + 15 = 15 + 47.962$

**6.** $23 - 6 = 6 - 23$

**7.** $\frac{1}{2} + \frac{1}{4} = \frac{1}{4} + \frac{1}{2}$

**8.** $4{,}972 + 15{,}237 = 15{,}237 + 4{,}972$

**9.** $3\frac{1}{2} \times 6\frac{1}{8} = 6\frac{1}{8} \times 3\frac{1}{2}$

**10.** $\$49.95 - \$23.72 = \$23.72 - \$49.95$

## B. Associative Property—Changing Groupings

When adding or multiplying three or more numbers, the way the numbers are grouped will not change their sum or product.

Parentheses are used to indicate which numbers are to be added or multiplied first.

Note that the Associative Property does not hold for subtraction or division.

**Simple Proof** for a balanced equation using the Associative Property:

$$(15 + 4) + 9 = 15 + (4 + 9)$$
$$19 + 9 = 15 + 13$$
$$28 = 28$$

$$(4 \cdot 5) \cdot 3 = 4 \cdot (5 \cdot 3)$$
$$20 \cdot 3 = 4 \cdot 15$$
$$60 = 60$$

Why the Associative Property does not work for subtraction and division:

$$(10 - 6) - 3 \neq 10 - (6 - 3)$$
$$4 - 3 \neq 10 - 3$$
$$1 \neq 7$$

$$(20 \div 5) \div 2 \neq 20 \div (5 \div 2)$$
$$4 \div 2 \neq 20 \div 2.5$$
$$2 \neq 8$$

**Exercise 4** Use the Associative Property to prove the equality (or in equality) in these problems.

**1.** $\left(\frac{1}{2}+\frac{1}{4}\right)+\frac{1}{8}=\frac{1}{2}+\left(\frac{1}{4}+\frac{1}{8}\right)$

**2.** $(17-6)-4=17-(6-4)$

**3.** $.52 \cdot (.4 \cdot .08) = (.52 \cdot .4) \cdot .08$

**4.** $(25+15)+60=25+(15+60)$

**5.** $5\frac{1}{8}\times\left(6\times 4\frac{1}{2}\right)=\left(5\frac{1}{8}\times 6\right)\times 4\frac{1}{2}$

**6.** $36 \div (6 \div 3) = (36 \div 6) \div 3$

**7.** $\left(1\frac{1}{2}+2.75\right)+1\frac{1}{4}=1\frac{1}{2}+\left(2.75+1\frac{1}{4}\right)$

**8.** $(\$238 \times 6) \times 4 = \$238 \times (6 \times 4)$

**9.** $(12\times 12) \times .12 = 12(12\times .12)$

**10.** $14.86 + (1.4 + .5) = (14.86 + 1.4) + .5$

## C. Distributive Property

The Distributive Property is used most frequently when multiplication is distributed to a sum or a difference.

In the following equation, **4** is distributed to the 3 and 5 before multiplication and addition are performed:

| | |
|---|---|
| $\mathbf{4}(3+5)=\mathbf{4}\times 3+\mathbf{4}\times 5$ | Distribute 4. |
| $4(3+5)=12+20$ | Multiply : $4\times 3$ and $4\times 5$. |
| $4(3+5)=32$ | Add : $12 + 20$. |

And in this equation, **4** is distributed to 6 and 3 before multiplication and subtraction are performed:

| | |
|---|---|
| $\mathbf{4}(6-3)=\mathbf{4}\times 6-\mathbf{4}\times 3$ | Distribute 4. |
| $4(6-3)=24-12$ | Multiply : $4\times 6$ and $4\times 3$. |
| $4(6-3)=12$ | Subtract : $24-12$. |

**Simple Proof**, using the Distributive Property, to determine if an equation is balanced:

$$3(10+2) = \underline{\qquad}$$

$$3(10+2)=(3\cdot 10)+(3\cdot 2) \qquad \text{Distribute 3.}$$

Add 10 + 2.  $\qquad 3(12) = 30+6 \qquad$ Multiply $3\times 10$ and $3\times 2$.

Multiply $3\times 12$.  $\qquad 36 = 36 \qquad$ Add 30 + 6.

Left side = Right side

**A.** Prove that the Distributive Property works for these equations. The first problem is done for you.

$$\textbf{1.}\ \ 4(10 + 2) = (4 \cdot 10) + (4 \cdot 2)$$
$$4(12) = 40 + 8$$
$$48 = 48$$

**2.** $5(4 - 2) = (5 \cdot 4) - (5 \cdot 2)$

**3.** $3(9 + 4) = (3 \cdot 9) + (3 \cdot 4)$

**4.** $2(6 - 3) = (2 \cdot 6) - (2 \cdot 3)$

**5.** $6(6 + 1) = (6 \cdot 6) + (6 \cdot 1)$

**6.** $3(20 - 15) = (3 \cdot 20) - (3 \cdot 15)$

**7.** $11(9 - 5) = (11 \cdot 9) - (11 \cdot 5)$

**8.** $12(7 + 6) = (12 \cdot 7) + (12 \cdot 6)$

**9.** $10(3 + 6) = (10 \cdot 3) + (10 \cdot 6)$

**B.** Apply the Distributive Property to these problems. The first one is done.

$$\textbf{1.}\ \ 3(4 - 2) = (\textbf{3} \cdot \textbf{4}) - (\textbf{3} \cdot \textbf{2})$$
$$3(2) = 12 - 6$$
$$6 = 6$$

**2.** $6(10 + 2) =$

**3.** $4(25 + 10) =$

**4.** $3(22 + 11) =$

**5.** $7(14 + 3) =$

**6.** $9(10 - 6) =$

**7.** $5(14 + 6) =$

**8.** $25(4 - 3) =$

**9.** $20(14 + 15) =$

**10.** $.5(22 - 19.1) =$

Review of Properties. Name the property as Associative, Commutative, or Distributive used in each problem.

**1.** $12 \cdot 6 = 6 \cdot 12$

**2.** $3(4 + 2) = (3 \cdot 4) + (3 \cdot 2)$

**3.** $(4 + 3) + 2 = 4 + (3 + 2)$

**4.** $(6 \cdot 5) \cdot 1 = 6 \cdot (5 \cdot 1)$

**5.** $(52 \ 6) \cdot 14 = 52 \cdot (6 \cdot 14)$

**6.** $27(19 - 15) = (27 \cdot 19) - (27 \cdot 15)$

**7.** $142 + 194 = 194 + 142$

**8.** $(3 + 4) + (5 + 6) = (5 + 6) + (3 + 4)$

**9.** $.75 \times .06 = .06 \times .75$

**10.** $127(15 + 142) = (127 \cdot 15) + (127 \cdot 142)$

**11.** $\left(\frac{1}{4} + \frac{1}{2}\right) + \frac{1}{8} = \frac{1}{4} + \left(\frac{1}{2} + \frac{1}{8}\right)$

**12.** $\frac{2}{3}(4 + 2) = \left(\frac{2}{3} \cdot 4\right) + \left(\frac{2}{3} \cdot 2\right)$

**13.** $12 \cdot 9 \cdot 4 = 4 \cdot 9 \cdot 12$

**14.** $7\left(4\frac{1}{2} - 3\frac{3}{8}\right) = \left(7 \cdot 4\frac{1}{2}\right) - \left(7 \cdot 3\frac{3}{8}\right)$

## D. Order of Operations

Without already being told, you have been using the Order of Operations to solve problems. The order in which operations are used is important to the outcome of a problem.

For instance, which solution is correct?

$$2 + 3 \cdot 4 = 2 + 12 = 14$$
or
$$2 + 3 \cdot 4 = 5 \cdot 4 = 20$$

When a problem has more than one operation, follow this order for a solution:

1. Do any operation with parentheses.
2. Solve all exponents ($3^2 = 9$).
3. Multiply and divide from left to right.
4. Add and subtract from left to right.

---

**Exercise 7**    Order of Operations.

**A.** Above, which answer is correct 14 or 20? Why?

**B.** Use the Order of Operations to solve these equations.

**1.** $(40 \cdot 2) - (6 \cdot 11) =$     **2.** $28 \div 2 + 4 =$     **3.** $6 \div .5 + 6 =$

**4.** $5 \cdot 3^2 - (15 - 6) =$     **5.** $(5 \div 1) + 6 =$     **6.** $(7 + 3) \div 5 =$

**7.** $9^2 + (4 + 6 - 9) =$     **8.** $12 \div 3 + 12 \div 4 =$     **9.** $36 - 5 \cdot 6 =$

**10.** $(18 - 4) \div 7 + 8 =$     **11.** $35 - 6 \cdot 4 =$     **12.** $3 + 4^3 + 27 \cdot 4 =$

**13.** $2 \cdot 6 - 5 + 2 =$     **14.** $(4 \cdot 6) + (25 \div 5) =$     **15.** $81 \div 9 - (5 - 3) =$

**16.** $2^2 + 3^2 - (12 - 2) =$     **17.** $(66 \div 11) + 4 - 5 =$     **18.** $(8 \cdot 8) - (9 \cdot 6) =$

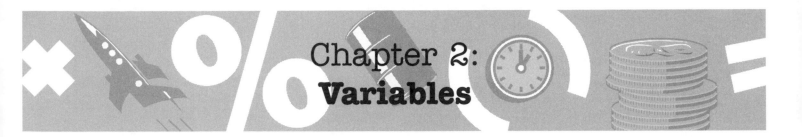

# Chapter 2: Variables

You are familiar with these types of equations:

$$12 + 6 = 18$$

$$12 + 6 = \underline{\ ?\ }$$

Here is an equation with something new, a letter:

$$12 + 6 = b$$

The letter ($b$) is a variable. A **variable** is a letter or symbol used to represent a missing number.

## Part 1 — Expressions and Terms

When variables are used with other numbers, parentheses, or operations, they create an algebraic **expression**.

| Algebraic Expressions | Regular Math Expressions |
|---|---|
| $a + 2$ | $3 \cdot 12$ |
| $3(a + 2)$ | $127 - 42 + 6$ |
| $(a)\ (b)$ | $6(3 + 2)$ |
| $3m + 6n - 6$ | $15 - 4 \cdot 6$ |

**Term** is the name given to the individual parts of an expression.

| Algegraic Expressions | Terms |
|---|---|
| $a + 2$ | $a, 2$ |
| $3(a + 2)$ | $3, a, 2$ |
| $(a)\ (b)$ | $a, b$ |
| $3m + 6n - 6$ | $3m, 6n, 6$ |

**A.** Use these terms and operations to make an algebraic expression.

**1.** $a$, $b$, +

**2.** $3r$, $2r$, −

**3.** $\frac{y}{2}$, $\cdot$ , 3

**4.** $2a$, $b$, +, ( )

**5.** 3, ( ), $2b$, +, 3

**6.** $r$, $3b$, $\cdot$

**7.** $6b$, $2b$, ÷

**8.** $a$, $(2 + 3)$

**9.** 16, − , $y$

**10.** 25, $c$, ÷

**11.** 3, $\cdot$ , $m$, +, 2

**12.** $24m$, +, $17m$

**13.** 8, $d$, +

**14.** $3b$, ( ), $6 \cdot 4$

**15.** $a$, $b$, $c$, +, ( )

**16.** The number 25 subtracted from $b$.

**17.** The number 365 times $g$.

**18.** The number 12 divided by $v$.

**19.** The variable $x$ added to 12.

**20.** The variable $x$ divided by the variable $x$.

**B.** What are the terms that make these expressions?

**1.** $6y + 10$

**2.** The variable $n$ added to 16.

**3.** $4a$

**4.** Twelve times $b$.

**5.** $12a - 6b + 4$

**6.** The number 4 divided by 2 plus 10.

**7.** $\frac{1}{2}ab + 3$

**8.** Thirty-two subtracted from $c$.

**9.** $4x - 2y$

**10.** Two plus 3 times $x$.

**C.** Match each word phrase with its equivalent algebraic expression.

____ **1.** The sum of a number and 10.

    **a.** $\frac{1}{2}a - 20$

____ **2.** The product of a number and 15.

    **b.** $\frac{a-4}{3}$

____ **3.** One half of a number minus 20.

    **c.** $a + 10$

____ **4.** A number decreased by 4 then divided by 3.

    **d.** $15a$

_____ **5.** Seven times a number plus 12.

_____ **6.** A number divided by the sum of 4 and 7.

_____ **7.** The quotient of 30 and 5 plus 10.

_____ **8.** Twenty-five divided by a number.

_____ **9.** Twice the sum of a number plus 4.

_____ **10.** A number added to a number.

_____ **11.** The difference between a number and 20.

_____ **12.** The sum of $\frac{3}{4}$ of a number and 7.

_____ **13.** The product of a number and 8 increased by 17.

_____ **14.** Nine less than twice a number.

_____ **15.** Ten times a number increased by 150.

**e.** $a + b$

**f.** $\frac{30}{5} + 10$

**g.** $8a + 17$

**h.** $10a + 150$

**i.** $\frac{a}{4+7}$

**j.** $2a - 9$

**k.** $\frac{25}{a}$

**l.** $2(a + 4)$

**m.** $7a + 12$

**n.** $\frac{3}{4}a + 7$

**o.** $a - 20$

**D.** Write word phrases for these expressions.

**1.** $2a + 4$

**2.** $50 - d$

**3.** $3(b + 3)$

**4.** $x + y + z$

**5.** $\frac{14-5}{c}$

**6.** $24r$

**7.** $\frac{9}{8}a - 2 + 4$

**8.** $8a + 9b$

**9.** $3(9a + 8b)$

**10.** $(14 - 12) + (3 + a)$

## Part 2 | Evaluating Expressions

If values are given for variables, algebraic expressions can be evaluated.

### ◆ Examples

- If $a = 6$, $b = 9$, and $y = 4$, evaluate $4(a - y)$:

  $4(a - y) =$

  $4(6 - 4) =$          Substitute values.

  $4(2) =$          Order of Operations: parentheses, multiplication.

  $8$          8 equals the value of $4(a - y)$.


- If $a = 6$, $b = 9$, and $y = 4$, evaluate $a + b - y$:

  $a + b - y =$

  $6 + 9 - 4 =$          Substitute values.

  $15 - 4$          $=$          Order of Operations: addition, subtraction.

  $11$          11 is the value of $a + b - y$.


### Exercise ② Evaluate these expressions.

Let $w = 7$, $x = 9$, $y = 3$, and $z = 1$.

**1.** $w \cdot y$          **2.** $x - y$          **3.** $9wy$

**4.** $2w + 2z + 2x$          **5.** $\frac{wx}{y}$          **6.** $4x - yx$

**7.** $xz \div y$          **8.** $(x - w)(y - z)$          **9.** $2x - 2w - z$

**10.** $yx - 20$          **11.** $5w - 3x + \frac{6y}{x}$          **12.** $138 \div y$

**13.** $(w + y) \div (x + z)$          **14.** $\frac{x+z}{5}$          **15.** $12(x - w)$

**16.** $50 - 3(x + y)$          **17.** $2(w - z) + 3(x - 3)$          **18.** $7x - 4w + \frac{7z}{7}$

# Combining Like Terms

Combining like terms, sometimes called "collecting terms," is an orderly process that helps to solve problems. In the process of combining like terms, the Commutative, Associative, and Distributive Properties, and well as the Order of Operations, are important.

**Like terms** are terms that have numbers or variables with the same exponents. Remember, any number or variable has an exponent of 1 if it does not have a higher exponent shown. Thus, 2 is actually $2^1$ and $a$ is actually $a^1$.

| **Like Terms** | | **Unlike Terms** | |
|---|---|---|---|
| $3x$, $-5x$ | Same variable. | $3x$, $5$ | Variable and a constant. |
| $12y$, $\frac{2}{3}y$ | Same variable. | $12y$, $\frac{2}{3}z$ | Different variables. |
| $5$, $6$ | Same constants. | $5$, $b$ | Variable and a constant. |
| $6x^2$, $10x^2$ | Same variables and exponents. | $6x^2$, $10x^3$ | Same variables, but different exponents. |

## ◆Examples

• Simplify $6(3x + 2)$ :

$$6(3x + 2) = (6 \cdot 3x) + (6 \cdot 2)$$
$$= 18x + 12$$

Use the Distributive Property.
Multiply to combine like terms.

• Simplify $7a + 6c + 7a + 8c$:

$$7a + 6c + 7a + 8c = 7a + 7a + 6c + 8c$$
$$= 14a + 14c$$

Use the Commutative Property.
Add to combine like terms.

Combining like terms allows for orderliness. These expressions have disorderly terms:

$$2x - 5 + 2x^2 \qquad 3m + 2a - n + 2 \qquad 10 - 12r + 3r^2d$$

Establishing orderliness will give:

$$2x^2 + 2x - 5 \qquad 2a + 3m - n + 2 \qquad 3r^2d - 12r + 10$$

Orderliness means arranging exponents from highest order downward, alphabetical order of variables, and placement of constants (sole numbers) last.

**Exercise** **3**    Simplify these expressions by combining like terms.

1. $2a + 10a + 4a$

2. $5y - 12y + 10y$

3. $25b + 35b + 5$

4. $27stu - 14stu$

5. $5(a + b)$

6. $7r + 6c + 9c - 3r$

7. $6e - 9e + 5 + 10e$

8. $21y + 15 - 16y + 23$

9. $15ab + 3 - 7ab + 23$

10. $9k - 13w + 7 + 21k$

11. $4x^2 + 3x - 7x^3 + 4$

12. $a^2b + ab^2 + 2a^2b$

13. $7x^2 + 12xy - 7xy + 4x^2$

14. $15 + x^2 + y + x^2$

15. $x^2 + x^2 + x^3 - x^3$

16. $15c^3d + 17c^3d - 9$

17. $3 + 3x^2y + 12x^2y + xy$

18. $mn^2 + mn^2 + m^2n$

19. $32k + 32 + 32m + m + k$

20. $8(3xy + 2x) - 3w$

# Chapter 3: Solving Basic Linear Equations

An **equation** is a mathematical statement that describes two equal values. The equal values are separated by an equal sign. When a variable (and sometimes more than one variable) is part of an equation, the value of the variable (or variables) must be found to make the equation complete and true.

**Linear equations** contain variables in which the exponent of each variable is 1.

| Linear Equations | Non-linear Equations |
|---|---|
| $5x - 1 = 4$ | $x^2 + x - 10 = 20$ |
| $b + 2 = 12$ | $3x^3 + 4x^2 - 2x + 7 = 182$ |
| $9a = 81$ | |
| $\frac{2}{3}a + 2 = 14$ | |

The exponent 1 is not commonly written, but understood—as mentioned in Chapter 2.

## Part 1  Isolating Variables

To solve a linear equation, isolate the variable to determine its value. Isolating the variable requires getting the variable alone on one side of the equation. Use the following four properties to help isolate variables.

## A. Addition Property

The same number can be added to both sides of an equal sign without changing the answer.

### ◆Example

Find the value of the variable ($c$) in the equation $c - 12 = 14$.

$$c - 12 = 14$$
$$c - 12 + \mathbf{12} = 14 + \mathbf{12}$$
$$c = 26$$

Isolate the variable $c$ by **adding 12** to both sides.

**Simple Proof**: Check the answer by substituting the value of $c$ into the original equation.

$$26 - 12 = 14$$
$$14 = 14$$

## B. Subtraction Property

The same number can be subtracted from both sides of an equal sign without changing the answer.

### 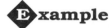xample

Find the value of the variable ($m$) in the equation $m + 47 = 198$.

| | |
|---|---|
| $m + 47 = 198$ | Isolate the variable $m$ |
| $m + 47 - \mathbf{47} = 198 - \mathbf{47}$ | by **subtracting 47** from both sides. |
| $m = 151$ | |

**Simple Proof:**

$$\mathbf{151} + 47 = 198$$
$$198 = 198$$

## C. Multiplication Property

The same non-zero number can be multiplied on both sides of an equal sign without changing the answer.

### 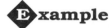xample

Find the value of the variable ($w$) in the equation $\frac{w}{4} = 2$.

| | |
|---|---|
| $\frac{w}{4} = 2$ | Isolate the variable $w$ |
| $\mathbf{4}\left(\frac{w}{4}\right) = \mathbf{4}(2)$ | by **multiplying** both sides by **4**. |
| $\frac{4w}{4} = 8$ | |
| $w = 8$ | |

**Simple Proof:**

$$\frac{\mathbf{8}}{4} = 2$$
$$2 = 2$$

## D. Division Property

The same non-zero number can be divided into both sides of an equal sign without changing the answer.

### ◆Example

Find the value of the variable ($x$) in the equation $9x = 72$.

$$9x = 72$$
$$\frac{9x}{9} = \frac{72}{9}$$
$$x = 8$$

Isolate the variable $x$
by **dividing** both sides by 9.

**Simple Proof:**
$$9 \cdot 8 = 72$$
$$72 = 72$$

### Did You Realize...?

Now that the four properties have been presented, did you realize that they have something to do with **inverse operations**?

• To solve an equation with an **addition** sign ($a + 10 = 32$), use **subtraction**.

• To solve an equation with a **subtraction** sign ($k - 32 = 41$), use **addition**.

• To solve an equation involving **multiplication** ($3r = 12$), use **division**.

• To solve an equation involving **division** ($\frac{x}{2} = 10$), use **multiplication**.

### Exercise ①  Solve for the variable in each equation.

**A.**

**1.** $a + 10 = 32$

**2.** $z + 8 = 43$

**3.** $52 + x = 104$

**4.** $w + 4.8 = 9.7$

**5.** $g + 3\frac{1}{2} = 11\frac{1}{4}$

**6.** $b + 12 = 38.5$

**7.** $157 + 126 = r$

**8.** $2.62 + v = 5.971$

**9.** $h + 129 = 736$

**10.** $c + \frac{3}{8} = 14\frac{1}{4}$

**B.**

**1.** $k - 32 = 41$  **2.** $15 = r - 6$

**3.** $f - 2.43 = 9.54$  **4.** $x - 6 = 14\frac{1}{2}$

**5.** $t - 6 = 5.62$  **6.** $p - \frac{3}{5} = \frac{3}{5}$

**7.** $y - 15 = 0$  **8.** $n - 4\frac{2}{3} = 6\frac{1}{6}$

**9.** $6.7 = d - 1.4$  **10.** $s - 129 = 482$

**C.**

**1.** $\frac{x}{2} = 10$  **2.** $\frac{b}{7} = 7$

**3.** $\frac{f}{20} = 3$  **4.** $\frac{a}{81} = .9$

**5.** $\frac{m}{11} = 5$  **6.** $\frac{d}{5} = 125$

**7.** $\frac{v}{6} = 7.05$  **8.** $\frac{e}{12} = 12$

**9.** $\frac{n}{30} = 5$  **10.** $\frac{n}{23} = 10$

**D.**

**1.** $3r = 39$  **2.** $1.5c = 9.3$

**3.** $\$2.76 = 12g$  **4.** $100w = 1234$

**5.** $.175d = .00175$  **6.** $1225 = 25q$

**7.** $5h = \frac{1}{2}$  **8.** $100j = 49$

**9.** $6x = 126$  **10.** $3\frac{1}{2}w = 42$

**E.**

**1.** $\$17.95 + m = \$52.75$  **2.** $f - 14\frac{7}{8} = 30\frac{1}{8}$

**3.** $\frac{2}{3} = 10b$  **4.** $\frac{1}{3}k = .62$

**5.** $\$124 = \frac{m}{4}$  **6.** $y - .375 = \frac{5}{8}$

**7.** $2.1 + c = 5$  **8.** $t + 76.884 = 101.5$

**9.** $\frac{h}{9} = \$.81$  **10.** $n \div 86 = 5$

# Equations from Word Problems and Phrases

Solving an equation is easiest if the equation is already provided:

$2x = 10$, find the variable $x$.

Translating words into an equation and then solving it is more difficult:

A number plus 10 is equal to 42.
$$x \quad + \quad 10 \quad = \quad 42$$

Turning words into expressions and expressions into equations takes practice.

**Phrase** ⟶ **Translation**

| Phrase | Translation |
|---|---|
| Six more than a number. | $x$ (the number) $+ 6$ |
| A number increased by 3. | $x + 3$ |
| Six more than a number is 10. | $x + 6 = 10$ |
| A number increased by 3 equals 7. | $x + 3 = 7$ |
| | |
| Four less than a number. | $x - 4$ |
| A number decreased by 9. | $x - 9$ |
| Four less than a number equals 7. | $x - 4 = 7$ |
| A number decreased by 9 is 27. | $x - 9 = 27$ |
| | |
| Twice a number. | $2x$ |
| Four times a number. | $4x$ |
| Twice a number is 28. | $2x = 28$ |
| Four times a number equals 16.4. | $4x = 16.4$ |
| | |
| One-third of a number. | |
| A number divided by 6. | $\frac{x}{6}$ |
| One-third of a number is 12. | $\frac{1}{3}x = 12$ |
| A number divided by 6 equals 42. | $\frac{x}{6} = 42$ |

## Exercise 2

**A.** Translate these phrases into expressions.  Use *n* for any unstated variable.

**1.** Five less than a number.

**2.** A number increased by 123.

**3.** Some number increased by 12.

**4.** Twenty-seven times some number.

**5.** A number minus .58.

**6.** Four times an unknown quantity.

**7.** One hundredth of a number.

**8.** A variable multiplied by 15.

**9.** Ten subtracted from a number.

**10.** The product of a variable and 10.

**11.** Fourteen divided by a number.

**12.** Ten decreased by y.

**13.** Forty-two increased by x.

**14.** Ninety-seven divided by 2.

**15.** The product of m and 12.

**16.** The sum of 132 and an unknown number.

**B.** Write an equation using the variable *c*.

**1.** The sum of a number (*c*) and 8 is 35.

**2.** Twice a number is 42.

**3.** Eighteeen less than a number is 172.

**4.** One-fourth of a number is 20.

**5.** A number increased by 100 equals 427.

**6.** Three times a number is 45.

**7.** Some number take away 32 leaves 47.

**8.** An unknown number divided by 20 is 7.5.

**C.** Solve each equation written for Part B.

**D.** Write an equation.  Solve the equation.

**1.** Sixteen boxes cost $56.96.  What does each box cost?
(Hint: Let *b* = cost of each box)

Write an equation: _____ b  = $_____

Solve equation: *b* = $_____

**2.** Three compact discs are on sale for $34.05.  What is the cost of each disc?  (Hint: Let $c$ = cost of each disc.)

**3.** Elena has $27.50.  She wants to make a purchase costing $35.20.  How much does she still need?  (Hint: Let $a$ = amount still needed.)

Write an equation: $_____ + a = $_____

Solve equation: $a = $_____

**4.** If 19.4 is added to a number, the sum is 98.5.  What is the number?  (Hint: Let $n$ = the number.)

**5.** Jane has $428 left in her bank account after spending $114.  How much did she have before she spent her money?  (Hint: Let $m$ = amount before spending $114.)

Write an equation: $m - _____ = $428$

Solve equation: $m = $_____

**6.** Miles bought several pieces of fruit for $1.29 and had $.71 left over.  How much money did he have before buying the fruit?  (Hint: Let $b$ = money before.)

**7.** Julio planted 36 bulbs, which is $\frac{1}{2}$ of the number of bulbs he bought.  How many bulbs did he buy?  (Hint: Let $t$ = total bulbs bought.)

Write an equation: $\frac{1}{2}$ ___ = _____

Solve equation: _____ = _____

**8.** M.T. Glass needs three times his allowance to buy a new coat.  The coat costs $37.50.  How much is M.T.'s allowance?  (Hint: Let $a$ = M.T.'s allowance.)

**9.** Jewel has picked 6.3 kilos of blueberries.  How many more kilos must Jewel pick to have a total of 12 kilos?  (Hint: Let $m$ = more kilos.)

**10.** One-third of our monthly income goes for the rent.  Rent is $528.  What is our monthly income? (Hint: Let $i$ = income.)

**11.** Two numbers have a sum of 187.  One number is 74.8.  What is the other number?  (Hint: Let $s$ = second number.)

**12.** My uncle is 39 years old. He is three times my age. How old am I? (Hint: Let $m$ = my age.)

**13.** A builder has 1334 bricks left after using 1366 bricks. How many bricks did the builder begin with? (Hint: Begin with a subtraction problem; let $b$ = beginning number of bricks.)

**14.** I weigh 125 pounds. John weighs 5 pounds less. How much does John weigh? (Hint: Begin with an addition equation; let $j$ = John's weight.)

**15.** One-third of our school is 136 students. How many students attend our school? (Hint: Let $t$ = total students.)

---

## Part 3 — Solving Two-step Linear Equations

So far only one operation has been used to solve an equation:

$$x + 2 = 15$$
$$x + 2 - 2 = 15 - 2 \qquad \text{Subtraction Property (inverse of addition).}$$
$$x = 13$$

The use of subtraction (the inverse of addition) is enough to isolate the variable $x$.

Since equations solve for unknowns (variables), more than one operation may be necessary to solve a linear equation.

### Examples

- Three times a number plus 2 equals 26. What is the number?

    Or, written as a word problem:

    I am 26 years old, which is three times Kayla's age plus two years. How old is Kayla?

$$3k + 2 = 26$$
$$3k + 2 - 2 = 26 - 2 \qquad \text{Subtraction Property to undo addition.}$$
$$3k = 24$$
$$\frac{3}{3}k = \frac{24}{3} \qquad \text{Division Property to undo multiplication.}$$
$$k = 8$$

Kayla is 8 years old.

- $\frac{2}{3}$ of a number subtract 2 equals 28. What is the original number?

Or, written as a word problem:

The cost of two-thirds of a yard, minus a $2.00 sales offer, equals $2.80. What was the original cost of one yard of fabric?

$$\frac{2}{3}c - \$2 = \$2.80$$

$$\frac{2}{3}c - \$2 + \mathbf{\$2} = \$2.80 + \mathbf{\$2} \qquad \text{Addition Property to undo subtraction.}$$

$$\frac{2}{3}c = \$4.80$$

$$\frac{2}{3}c \times \mathbf{\frac{3}{2}} = \$4.80 \times \mathbf{\frac{3}{2}} \qquad \text{Multiplication Property to undo division.}$$

$$c = \$7.20$$

Original cost of one yard of fabric was $7.20.

In Chapter 1, you learned about the Order of Operations. The Order of Operations allowed us to solve problems using more than one operation. You will notice that when you solve two-step linear equations, the Order of Operations you learned earlier is reversed. In the two preceding examples, addition and subtraction were performed before multiplication or division.

## Exercise ❸

**A.** Solve for the variable in each two-step equation.

**1.** $20 = 6x + 8$      **2.** $\frac{1}{2}x + \frac{1}{3} = \frac{2}{3}$      **3.** $0.3x - 8.4 = 7.2$

**4.** $7x - 6 = 78$      **5.** $\frac{3}{4}x - 9 = 27$      **6.** $3x + 5 = 26$

**7.** $5x + 3 = 28$      **8.** $4x - 7 = 25$      **9.** $\frac{5}{7}x - 13 = 83$

**10.** $\frac{x}{2} + 13 = 17$      **11.** $14x - 22 = 160$      **12.** $1.5x + 0.6 = 12.6$

**13.** $\frac{2x}{3} + 30 = 34$      **14.** $\frac{3}{4}x + \frac{2}{3} = 1\frac{1}{3}$      **15.** $28 - x = 8$
(Hint: Add $x$ to both sides.)

**B.** Combine terms before solving these equations.

   **1.** $3x + 5x - 2 = 22$                   **2.** $10x - 25 - 3x = 24$

   **3.** $12x + 23 - 7x = 98$             **4.** $4 - 3x + 11 + 9x = 45$

   **5.** $16.9 = x + 4.6 + 2x$             **6.** $23x + 18 - 12x = 62$

   **7.** $33x - 21 - 9x = 171$           **8.** $\frac{1}{2}x + \frac{1}{2} - \frac{1}{4}x = 3$

   **9.** $29x + 33 + 14x = 205$        **10.** $3\frac{3}{8} + 4\frac{1}{4}x - 2\frac{1}{4}x = 5\frac{1}{4}$

  **11.** $17 + 3x - 14 - x = 21$        **12.** $55x + 21 + 17x = 165$

**C.** Distribute, collect, and solve these equations.

   **1.** $4(2x + 7) = 108$               **2.** $5(3x + 8) = 175$

   **3.** $(3x + 2) + (9x - 1) = 37$       **4.** $2(3x + 14) - 28 = 27$

   **5.** $3x + 2(5x - 3) = 7$           **6.** $6(2x - 2) + 8 = 32$

   **7.** $3x + 4(3x - 5) = 25$         **8.** $2(2x + 7) + (3x - 5) = 37$

   **9.** $13 + 8(5x - 9) = 21$         **10.** $(12x + 30) - 2(4x) = 90$

  **11.** $(2x - 7) + (7x + 10) = 84$    **12.** $6(5x - 9) + 24 = 120$

At this point you might realize a general procedure to solve equations:

   1. Use the Distributive Property to remove parentheses when necessary.

   2. Combine like terms.

   3. Use the Order of Operations in reverse to solve for the variable (addition/subtraction before multiplication/division).

An equation can have a variable on both sides of the equal sign:

$$5b = 15 + 2b$$

Always isolate the variable to one side or the other. Perform inverse operations until the value of the variable is determined.

| | |
|---|---|
| $5b = 15 + 2b$ | Isolate the variable by performing the inverse of addition, the Subtraction Property. |
| $5b - \mathbf{2b} = 15 + 2b - \mathbf{2b}$ | |
| $3b = 15$ | Combine terms. |
| $\frac{3}{3}b = \frac{15}{3}$ | Perform the inverse of multiplication by using the Division Property. |
| $b = 5$ | |

A variable can be isolated on either side of an equation. Above, the variable was isolated on the left side. Had it been isolated on the right side, more complex operations would be necessary (working with negative numbers), but the answer would be the same. Isolating a variable to the left of the equal sign seems most common.

## ◆xamples

- Here is an example that begins with the Distributive Property:

| | |
|---|---|
| $3(a + 8) = 7a$ | |
| $\mathbf{3a + 24} = 7a$ | Distributive Property. |
| $3a - \mathbf{3a} + 24 = 7a - \mathbf{3a}$ | Subtraction Property. |
| $24 = 4a$ | Combine terms. |
| $6 = a$ | Division Property. |

- If you are familiar with negative integers, the variable could have been isolated on the left side in this fashion:

| | |
|---|---|
| $3(a + 8) = 7a$ | |
| $\mathbf{3a + 24} = 7a$ | Distributive Property. |
| $3a - \mathbf{7a} + 24 = 7a - \mathbf{7a}$ | Subtraction Property. |
| $-4a + 24 = 0$ | Combine terms. |
| $-4a + 24 - \mathbf{24} = 0 - \mathbf{24}$ | Subtraction Property. |
| $-4a = -24$ | Combine terms. |
| $a = 6$ | Division Property. |

**1.** $12g = 45 - 3g$

**2.** $5x + 16 = 9x$

**3.** $5(v + 6) = 8v$

**4.** $6y = y + 25$

**5.** $11z = 28 + 71$

**6.** $5w + 4 = 2w + 16$

**7.** $12s - 4 = 8s$

**8.** $2m = 5m - 3$

**9.** $2c + 16 = 5c + 4$

**10.** $3k = 13k - 20$

**11.** $12x + x = 5 - 2x$

**12.** $6d - 14 = 2d + 6$

**13.** $3f + 30 = 8f$

**14.** $8e - e = e + 24$

**15.** $3(n + 5) = 8n$

**16.** $4\left(\frac{x}{4} + 3\right) = 12$

**17.** $9(8a - 7) = 153$

**18.** $44 - 8r = -4r$

**19.** $2(k + 7) = 9k$

**20.** $-w - 2 = 1 - 2w$

**21.** $45 = 3(a + 5)$

**22.** $8(4 - x) = 5(2 - x)$

**23.** $9(c - 4) = 3(c + 12)$

**24.** $4(p + 5) = 2(p + 40)$

**25.** $8n + 6 = 5(n + 12)$

**26.** $5c - \frac{1}{4} = 4c + \frac{3}{4}$

**27.** $2(5p - 7) = 9p + 17$

**28.** $8(3b + 1) = 7(2b + 4)$

**29.** $3(2x + 4) = 2(9 + 2x)$

**30.** $2(11) = 2(y + 4)$

# Chapter 4:
# Word Problems

You worked with word problems in the last chapter. You know that words must be translated (changed) into expressions before an equation can be written and then solved.

## Part 1 | Simple Word Problems

From Chapter 3, you are familiar with simple word problems of this type:

Eight times a number is 40. Find the number.

Translate:    $8 \cdot n = 40$          $n$ = unknown number.
              $8n = 40$         Apply Division Property.
               $n = 5$

 **Exercise 1**   Translate these simple word problems into equations. Use $n$ as a variable. Solve the equations.

**1.** Four more than a number is 15. Find the number.

**2.** Forty-five less than a number is 82. Find the number.

**3.** Seven times a number equals 77. Find the number.

**4.** One-half of a number is $14\frac{1}{2}$. Find the number.

**5.** Twice a number divided by 6 is 42. Find the number.

**6.** Forty-five divided by a number is 5. Find the number.

**7.** Twelve more than a number divided by 3 is 9. Find the number.

**8.** Four times two less than a number equals 14. Find the number.

**9.** The sum of four and twice a number equals 52. Find the number.

**10.** If five times a number is decreased by 12, the answer is 9. Find the number.

# More Complex Word Problems

Simple word problems have few words. But when real-life names, places, and situations are included, word problems become more complex.

To help solve more complex word problems, use these Five Steps:

1. Read (again and again, if necessary) the word problem to determine what must be solved.

2. Identify the unknown quantity (sometimes quantities) and assign a variable.

3. Translate into expression(s) forming an equation.

4. Isolate the variable to one side.

5. Solve the equation.

## ◆xamples

- A student had $257.75 in her savings account before making a $32.95 withdrawal. What is the balance after the withdrawal?

  1. What to solve:     Balance after withdrawal.

  2. Assign a variable:     Let $b$ = balance after withdrawal.

  3. Translate:     $257.75 − $32.95 = $b$

  4. Isolate variable:     Already isolated.

  5. Solve:     $b$ = $224.80.

- Later, the same student had a balance of $627.83. The student makes a deposit of $27.42 and a withdrawal of $106.09. What is the new balance?

  1. What to solve:     New balance.

  2. Assign a variable:     Let $n$ = new balance.

  3. Translate:     $627.83 + $27.42 − $106.09 = $n$

  4. Isolate variable:     Already isolated.

  5. Solve:     $n$ = $549.16

• One-twelfth of our bulbs are tulips. That's 17 tulip bulbs. How many total bulbs do we have?

    1. What to solve:    Total number of bulbs.

    2. Assign a variable:    Let $t$ = total number of bulbs.

    3. Translate:    $\frac{1}{12}t = 17$

    4. Isolate variable:    $t = 17 \cdot 12$

    5. Solve:    $t = 204$.

• The community theatre sold 257 tickets for the Friday performance and twice as many plus 50 for the two Saturday performances. How many tickets were sold for the Saturday performances?

    1. What to solve:    Number of Saturday tickets sold.

    2. Assign a variable:    Let $s$ = Saturday tickets sold.

    3. Translate:    $2(257) + 50 = s$

    4. Isolate:    Already isolated.

    5. Solve:    $s = 564$.

**Exercise**  ❷    For the following problems, write an equation with the variable isolated on the left and solve the equation.

**1.** Four chairs to complement a new dining table cost $211. What is the cost of each chair?

**2.** If our club collects 792 more cans of food, it will have a total of 15,000 cans. How many cans have already been collected?

**3.** Provincetown is 47 km from here. North Pence is twice as far, less 12 km. How far is North Pence from here?

**4.** Milk, bread, and fruit cost $15.20 at the market. Milk was $2.59, and two loaves of bread together were $4.30. How much did the fruit cost?

**5.** A company's daily payroll just for its twenty-three electricians is $3513.02. If each electrician receives the same pay, how much does each earn daily?

**6.** Derrick was born in 1986. How old will he be in 2009?

**7.** Marta is 6 inches taller than her brother.  Find her brother's height if Marta is 72 inches tall.

**8.** During winter, monthly natural gas costs can triple compared to the combined costs of water and electricity.  If in January the Singh family pays $21.50 for electricity and $14.50 for water, what might they expect to pay for natural gas?

**9.** I need four times my savings to purchase an air conditioner.  The air conditioner costs $695.  What is my present savings?

**10.** Paula saves one-seventh of her weekly salary.  That savings amounts to $62 per week.  How much is her weekly salary?

**11.** If $\frac{3}{4}$ pound of meat costs $3.90, what does one pound cost?

**12.** A new car sells for $12,500, which is $500 more than 15 times the cost of a canoe.  What is the cost of a canoe?

**13.** Jane is six times older than her granddaughter. If Jane is 66, how old is her granddaughter?

**14.** A rancher has a 1350-acre farm.  This year she will lease 360 acres and graze cattle on 720 acres.  How many acres will not be used?

**15.** If Sarah was driving a race car at 200 kilometers per hour, what was her speed in miles per hour?  (1 mile = 1.6 kilometers)

**16.** If Tiberius was born on the first day of 42 BC and died on the second day of 37 AD, how old was he when he died?  (Remember difference in BC/AD.)

**17.** A one-liter carton of milk is $1.25.  A two-liter carton is $2.25.  If Emily buys 8 liters in two-liter cartons, how much will she save as compared to buying 8 liters in one-liter cartons?

**18.** Lin needs to balance her checkbook after shopping on Saturday.  She spent $64.35 for groceries, $12.95 for a birthday gift, $147.20 on family clothing, and $24.95 for several CDs.  How much did she spend on Saturday?

Lin's balance before shopping was $892.57.  What was her bank balance after shopping?

**19.** Cal's gas tank is $\frac{1}{4}$ full.  The tank holds 30 liters.
a. How many liters are in the gas tank now?
b. How many liters are needed to fill the tank?
c. How many liters must be added to make the tank $\frac{2}{3}$ full?

# Part 3

# One Variable to Solve Several Unknowns

All previous work has required you to solve for one variable, such as:

Nine less than a number is 35. What is the number?

1. What to solve:    Number.

2. Assign a variable:    Let $n$ = number.

3. Translate:    $n - 9 = 35$

4. Isolate:    $n = 35 + 9$

5. Solve:    $n = 44$.

Word problems then become more complex by adding more information:

So-and-so had a beginning bank balance of \$35. He bought a blah-blah for \$9. What is So-and-so's balance now?

1. What to solve:    Balance now.

2. Assign a variable:    Let $n$ = new blance.

3. Translate:    $n = \$35 - \$9$

4. Isolate:    Already isolated.

5. Solve:    $n = \$26$.

Now, how can you handle **two unknowns** with one variable?

### ◆ Example

Maria is 3 years older than Elena. If the sum of their ages is 17, find the age of each girl.

Remember the Five Steps—

1. What to solve:    Age of each girl (two answers, not one).

2. Assign a variable:    But there are two girls. Use one variable to describe each girl.

   Let $E$ = Elena's age

   Let $E + 3$ = Maria's age (She is 3 years older than Elena.)

3. Translate:    $E + (E + 3) = 17$

4. Isolate:     $2E + 3 = 17$                 Combine terms.

$$2E + 3 - \mathbf{3} = 17 - \mathbf{3}$$       Apply Subtraction Property.

$$\frac{2E}{2} = \frac{17-3}{2}$$       Apply Division Property.

$$E = \frac{14}{2}$$

5. Solve:     $E = 7$
$$E + 3 = 10$$

Elena is 7 years old.
Maria is 3 years older ($E + 3$) or 10 years old.

---

## Exercise ❸   Use one variable to solve two unknowns.

1. One of two numbers is 26 more than the other.  The sum is 80.  Find the numbers.

2. Jessie has $26 more than Sadie.  Together they have $80.  How much does each have?

3. One number is 12 less than another.  Together both numbers total 30. What is each number?

4. Julius must separate 38 oranges into two bags.  Onebag must have 12 more oranges than the other.  How many oranges will be in each bag?

5. The sum of two numbers is 41.  The larger number is one less than twice the smaller number.  Find the numbers.  (Hint: One number is $2n - 1$)

6. Separate 90 into two parts.  One part must be four times the other.

7. Ida Claire is four times as old as Clara Fie.  The sum of their ages is 65 years.  Find the age of each.

8. One large container weighs 2.5 kg less than the other.  The weight of both together is121.5 kg.  What is the weight of each container?

9. In an office of 33 people, there are 7 more women than men.  How many men and women are there?

10. The sum of two numbers is 32.  The second number is eight more than 7 times the first.  What are the two numbers?

**11.** Ann Soforth practiced her piano 15 minutes longer on Sunday than on Saturday. If she practiced a total of 41 minutes this weekend, how long did she practice each day?

**12.** A plank is 5 m long. Eileen Slightly cuts it into two pieces so that one piece is 1 m less than the length of the other piece. What is the length of each piece?

**13.** Rex Itall receives 1.5 times his regular hourly wage for overtime work. He earned $315 last week for working 42 regular hours and 7 hours overtime. What is his regular hourly rate and his overtime rate? (Hint: One unknown is $1.5n \times 7$.)

**14.** A crowd of 91 people was divided into two groups. One group had 13 more people than the other. How many were in each group?

**15.** The sum of a number and 5 times that number is 18. What are the two numbers?

## Part 4 | Classic Word Problems

You have already worked with the classic word problems that require you to determine certain ages or to find certain numbers. This section will build upon those classic word problems to introduce consecutive number problems, coin and money problems, and geometric figure problems.

### A. Consecutive Number Problems

Numbers that succeed each other in normal order are consecutive. For example: 25, 26, 27 are consecutive numbers. The difference between consecutive numbers is one.

◆ **Example**

Find two consecutive numbers that total 35.

1. What to solve:   Two consecutive numbers that total 35.

2. Assign a variable:   $c$ = first number
$c + 1$ = second number

3. Translate:   $c + (c + 1) = 35$

4. Isolate:   $2c = 34$   Combine terms, Subtraction Property.

5. Solve:     $c = 17$          Division Property
            $c + 1 = 18$

One number = 17.
The other number = 18.


A slight variation is to find two consecutive odd numbers. (The difference is 2.)

## ◆xample

Find two consecutive odd numbers that total 36.

1. What to solve:    Two consecutive odd numbers that total 36.

2. Assign a variable:     $c$ = first number
                          $c + 2$ = second number

3. Translate:      $c + (c + 2) = 36$

4. Isolate:      $2c = 34$     Combine terms,  Subtraction Property.

5. Solve:     $c = 17$          Division Property.
            $c + 2 = 19$

One number = 17.
The other number = 19.


## B. Money Problems

Money problems can ask "how many things" (coins, bills) and/or "how much money" (value).  As you learned in consecutive number problems, defining the variables is the key to solving money problems.

## ◆xample

Felice had three times as many nickels as dimes.  If the total value of her coins was $1, how many of each coin did she have and what were their values?

1. What to solve:    a. Number of nickels, number of dimes.
                     b. Value of nickels, value of dimes.

2. Assign a variable:

|         | Number of coins | Value in dollars |
|---------|-----------------|------------------|
| Dimes   | $x$             | $.10(x)$         |
| Nickels | $3x$            | $.05(3x)$        |

3. Translate: Both sides of the equation must represent either the total number of coins or the total value of the coins. Because the problem des not indicate the total number of coins but does indicate their value as $1, both sides of the equation must represent the total value in dollar amounts.

$$.10x + .05(3x) = \$1.00$$

4. Isolate: $10x + .15x = 1.00$
$.25x = 1.00$      Combine terms.

5. Solve: $x = 4$      Division Property

    a. $x = \mathbf{4}$ = Number of dimes.
         $3x = \mathbf{12}$ = Number of nickels.

**Simple Proof:**
$4(.10) + 12(.05) = 1.00$
$\$.40 + \$.60 = \$1.00$

    b. $\mathbf{\$.40}$ = Value in dimes.
         $\mathbf{\$.60}$ = Value in nickels.

Variations on this type of coin problem may require a solution for only the number of each type of coin or only the total value for each type of coin. Related problems that involve, say, stamps, bills, or items with a set value (cost in dollar amounts) can also be solved with this "money" method.

## C. Geometric Figure Problems

Recall these facts about rectangles, squares, and triangles:

1. The perimeter of a rectangle equals 2 lengths + 2 widths:
     $P = 2l + 2w$      (We will discuss formulas in the next chapter.)

2. The area of a rectangle equals length × width:
     $A = l \times w$

3. A square has four equal sides:
     $P = 4s$          $A = s^2$

4. The sum of the three angles of any triangle equals 180°.
     $\angle A + \angle B + \angle C = 180°$

## ◆Example

The length of a certain rectangle is equal to twice the width. The perimeter is 138 units. What is the width of the rectangle?

1. What to solve:    Width of rectangle.

2. Assign a variable:    $w$ = width of rectangle
    $2w$ = length of rectangle

3. Translate:    Remember the formula for the perimeter of a rectangle: $P = 2w + 2l$
    $138 = 2w + 2(2w)$.

4. Isolate:    $138 = 6w$    Combine terms.

5. Solve:    $23 = w$    Division Property.
    $46 = 2w$

   Width = 23 units.
   Length = 46 units.

**Simple Proof**: Substitute the width and length values into the known formula:
$$P = 2w + 2l$$
$$138 = 2(23) + 2(46)$$
$$138 = 46 + 92$$
$$138 = 138$$

---

## Exercise ◆4  Consecutive Numbers, Money, and Geometric Figures.

**A.** Consecutive Number Problems. Solve these problems.

**1.** The sum of two consecutive numbers is 285. What are the numbers?

**2.** Find two consecutive even numbers with a sum of 86.

**3.** The sum of 3 consecutive numbers is 138. Find the numbers. (Hint: One number alone is $c$; a second number is $c + 1$; and the third number will be $c + ?$)

**4.** The sum of three consecutive odd numbers is 111. What are the numbers?

**5.** The sum of four consecutive numbers is 262. What are they?

**B.** Money Problems.  Solve these problems.

1. Juan has $.85 in change.  He has only nickels and dimes, twelve coins in all.  How many nickels and how many dimes does he have?  And what is the value of the nickels and the value of the dimes?  (Hint: Let $12 - d$ = number of nickels.)

2. An equal number of nickels, dimes, and quarters has a value of $2.00.  How many of each coin are there?

3. A coin collection of nickels and dimes is valued at $8.50.  How many coins are there?  There are three times as many nickles as dimes.

4. Rosetta Stone has five times as many quarters as dimes.  The total value of her coins is $16.20.  How many of each coin does she have and what is the total value of quarters?

5. A collection of coins is valued at $.64.  There are two more nickels than dimes and three times more pennies than dimes.  How many of each coin are there?  (Hint: Define the variables in terms of dimes.)

**C.** Geometric Figure Problems.  Solve these problems.

1. A rectangle has a length that is 4 meters less than 3 times the width.  The perimeter is 224 m.  What are the length and width?

2. A square and an equilateral triangle (all sides are equal) have the same perimeter.  Each side of the triangle is 8 meters.  Find the length of each side of the square.

3. The width of a rectangle is 3 cm less than $\frac{1}{2}$ its length.  The perimeter is 60 cm.  Find the length and width of the rectangle.

4. A landowner wants to fence a triangular parcel of land.  One side is twice the second; the third side is 20 m less than 3 times the second side.  The perimeter of the parcel is 106 m.  What is the length of each side?  (Hint: Use the second side as the basis for the variables.)

5. The second angle of a triangle is 20° less than the first.  The third is twice the second.  Find the third angle. (Hint: Let the third angle = $2(a - 20°)$.)

**D.** Solve these problems.

1. The first side of a triangle is 2 meters shorter than the second side. The third side is 5 meters longer than the second side. The perimeter of the triangle is 33 meters. What is the length of each side?

2. A piggy bank contained $12.25 in quarters, dimes, and nickels. The number of dimes was five more than twice the number of nickels. The number of quarters was less than three times the number of nickels. How many of each kind of coin were there?

3. The sum of three consecutive numbers is equal to six times the first number. What are the three numbers? (Hint: Let $n$ = first number, $n +$ ____ + ____ $= 6n$.)

4. The length of a rectangle is three times its width. If the perimeter of the rectangle is 24 feet, what is the area?

5. Three consecutive multiples of 5 total 120. What are the three multiples?

6. Guiditta has a number of nickels, twice as many dimes as nickels, and twice as many quarters as dimes. If she has 70 coins in all, what is the value of her dimes? How much money does she have altogether?

7. Two angles of a triangle have the same measure. The third angle is twice that of each of the other two. Find the measure of all three angles.

8. There are three consecutive numbers. The sum of the first two numbers is 35 more than the third number. What are the numbers? (Hint: Let $n$ = first number.)

9. Phil Atelist bought $30.64 worth of stamps. He bought twenty more $.19 stamps than $.50 stamps. And he bought twice as many $.29 stamps as $.19 stamps. How many of each kind did he buy? (Hint: Fill in the table below to help.)

| | Number of stamps | Value of stamps |
|---|---|---|
| $.50 | $n$ | $.50($n$) |
| $.19 | ____ | ____ |
| $.29 | ____ | ____ |

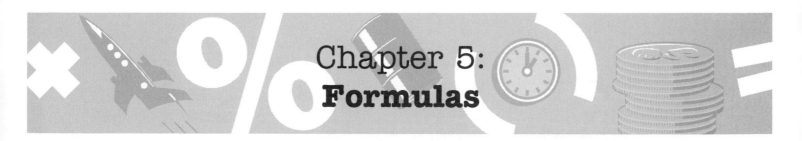

# Chapter 5: Formulas

A **formula** is an equation, usually applied to real life, that shows a relationship between two or more quantities represented by variables. All formulas have at least two variables.

## Common Formulas

| | |
|---|---|
| $A = bh$ | Area of a parallelogram |
| $V = lwh$ | Volume of a rectangular solid |
| $I = PRT$ | Calculation of simple interest |
| $C = \frac{5}{9}(F - 32)$ | Conversion of Fahrenheit to Centigrade |
| $C = \pi d$ | Circumference of a circle |
| $D = rt$ | Calculation of distance |

## Part 1: Solving for a Particular Variable

With a particular formula, it may be necessary to express a particular variable in terms of the other variables. An example is the standard interest formula, $I = PRT$, which stands for Interest = Principal x Rate x Times:

### ◆xample

Scrooge deposited $10,500 ($P$) into a savings account. At the end of one year ($T$), the interest earned ($I$) was $525. Find the interest rate ($R$).

1. What to solve:  Rate ($R$).

2. Assign a variable:  Standard formula, $I = PRT$.

3. Translate:  $525 = $10,500 ($R$) (1)

4. Isolate:  $R = \frac{I}{PT}$  Division Property.

5. Solve:  $R = \frac{525}{10,500\,(1)}$

$R = .05 = 5\%$

Interest rate = 5%.

## Exercise 1

**A.** Use your knowledge of properties to solve for the variable indicated.

**1.** $A = lw$; solve for $w$.  **2.** $I = PRT$; $P$  **3.** $D = rt$; $r$

**4.** $V = lwh$; $l$  **5.** $A = \frac{bh}{2}$; $b$  **6.** $F = \frac{9}{5}C + 32$; $C$

**7.** $V = \pi r^2 h$; $h$  **8.** $C = 2\pi r$; $r$  **9.** $P = 2l + 2w$; $l$

**10.** $C = \pi d$; $d$  **11.** $P = 4s$; $s$  **12.** $V = \frac{1}{3}bh$; $b$

**13.** $c^2 = a^2 + b^2$; $a^2$  **14.** $A = \frac{bh}{2}$; $h$  **15.** $A = lw$; $l$

**B.** Solve these word problems.

**1.** A sail is shaped in a triangle. The area of the sail is 160 square units. The height is known to be 20 units. Find the base of the sail. ($A = \frac{bh}{2}$).

**2.** Natalie Attired has an appointment in Portland, 135 miles away. Traveling at 60 mph, how quickly might she be able to travel to Portland? ($D = rt$)

**3.** A farmer must build a grain silo to store her grain harvest. She needs to store grain in units of at least 8478 cubic feet. Because of limited space, the diameter of the silo can be only 20 feet. How tall must the silo be? ($V = \pi r^2 h$, $\pi = 3.14$)

**4.** Kay Pasa just bought 319.8 square yards of carpeting. She remembered that the length of her room to be carpeted was 20.5 yards. What was the width? ($A = lw$)

**5.** How many years does it take to earn $564 interest on a $4700 savings account at 6% per year? ($I = PRT$)

**6.** A road builder has ordered 4810 cubic yards of fill for a hole that is 20 yards wide and 40 yards long. How deep must the hole be? Has the builder ordered enough fill? ($V = lwh$)

**7.** A major brand of paint will cover 420 square feet per gallon. A bedroom is 8 feet high. How many linear feet of wall can be painted with one gallon? ($A = lw$)

**8.** A circular hot tub at a fitness spa is 78.5 feet in circumference. What is the diameter of the tub? ($C = \pi d$, $\pi = 3.14$)

**9.** A carpenter must make flooring for a rectangular podium. She is told that the area of the podium must be 945 square meters. The longest side must be 35 meters. What must the width be? ($A = lw$)

**10.** A large cone with a diameter of 10 meters is part of an outdoor sculpture. The cone has been filled with a hardening foam for stability. It took 455.3 cubic meters of foam to fill the cone. What is the height of the cone? ($V = \frac{1}{3}\pi r^2 h$)

**11.** The perimeter of a rectangular pastureland is 919 feet. The length of the pastureland is 327.5 feet. What is the width? ($P = 2w + 2l$)

**12.** A cracker maker must design a box that holds 168 cubic inches of crackers. The cracker maker knows that the box must be 12 inches long and 8 inches wide. How thick must the box be? ($V = lwh$, thickness = $h$)

---

## Part 2 | Writing and Solving Formulas

At the very beginning of this book, you wrote simple equations for simple computational problems:

**Math Problem** ➤ **to** ➤ **Equation**

$$
\begin{array}{r}
152 \\
+\ 17 \\
\hline
169
\end{array}
$$

$$152 + 17 = 169$$

You are now at a point where you can write an equation to create a formula for a word problem.

### ◆**E**xample

A car is driven 910 miles and uses 40 gallons of gasoline. How many miles per gallon did the car get?

    1. What to solve:     Miles per gallon.

    2. Assign a variable:     $m$ = miles driven = 910
                              $g$ = gallons = 40
                              $x$ = miles per gallon = ?

    3. Translate:     $x = \frac{m}{g}$     Equation.

$$x = \frac{910}{40}$$

4. Isolate:   Already isolated.

5. Solve:   $x = 22.75$

22.75 miles per gallon

Use this model to write formulas. Be sure to assign each variable—unknown and known— a letter value, as in Step 2.

 **Exercise 2**   Write formulas and solve the following word problems.

**1.** On another trip the same car (as above) got 23.2 miles per gallon, using 32.5 gallons of gasoline. How many miles was the car driven?

**2.** The Farquart family are driving fools. On a recent trip, they covered 300 miles and used 15 gallons of gasoline. The average cost of gasoline was $1.20 per gallon. What was the gasoline cost per mile driven?

**3.** After their last trip, the Farquart family computed their costs and determined the cost per mile driven. But then they lost the calculations for the exact number of miles driven. They knew the average cost of gas was $1.25. They bought 10 gallons of gas and the cost per mile was $.08. How far had they driven?

**4.** Ella Gant purchased a winter coat which originally cost $129.95 at a 20% discount. How much did she save?

**5.** On her next purchase, Ella saved $4.08 on a pair of gloves originally costing $25.50. What percent discount did she receive on the purchase of gloves?

**6.** Anna Rexia is a sharp food shopper. She always compares unit prices of products. A large box of breakfast cereal costs $6.44. Its weight is 2 pounds .45 ounces (920 grams). What is the cost per gram?

**7.** With a Food Value Coupon, the unit price drops to $.006 per gram. How much does the box of cereal now cost?

**8.** The perimeter of an irregular lot having five sides is 398 meters. The lot's owner is told the measurements of four sides (27.5 m, 132 m, 29.5 m, and 133 m). What is the length of the fifth side?

**9.** Ann Soforth is paid to make telephone calls to remind patients of their appointments. She is paid $.17 per call and averages 527 calls per week. She is also paid a straight salary for other work. How much money does she average per month for phoning?

**10.** A small, local business has 12 employees. Monthly payroll is $17,131.44. What does the average worker earn per month?

**11.** What is the average yearly salary at the small business in Problem 10?

**12.** The business puts aside an amount equal to 12% of each employee's salary as a retirement fund. How much must the business put aside yearly for its employee retirement fund?

**13.** A restaurant advertises,"Buy one meal, pay half price for the second." A couple each orders a $12.50 meal. How much will the two meals cost, according to the advertisement?

**14.** A lot is 35 meters by 28 meters. How much would a fence cost to construct at $51.00 per meter?

**15.** A sales representative receives a 12% commission rate on all monthly sales. If the sales representative sells $24,572.27 in merchandise, what is her commission for the month?

**16.** Phyllis Teen's expensive watch gains 6 minutes every hour. She set the watch to the correct time at 9:00 AM. Later that day, Phyllis looks at her watch and notes that the time is 6:30 PM. What time is it really? (Solution involves 2 formulas: time gained, exact time.)

**17.** Larry spent 60% of his monthly salary on auto insurance. He spent $500, or $\frac{1}{5}$ of his salary, on groceries and meals. What is Larry's auto insurance cost? (Solution involves 2 formulas: salary, insurance.)

**18.** Owen Moore found that he paid 30% of his $40,000 annual income to income tax. How much less would Owen pay in income taxes if he moved to a location where the tax rate on his income was 10%?

**19.** Season tickets to hometown sporting events cost $255 per season. Each individual sporting event costs $15. What minimum number of games must be attended to make the season ticket worthwhile?

**20.** Flora Dation bought gloves for every member of her family as a Christmas gift. She has 12 family members and each pair of gloves costs $15, plus a 14% sales tax. How much money did Flora spend on gloves? (Solution involves 2 formulas: gloves, gloves + tax.)

# Answers

## Chapter 1: Equations

**Pages 6-7, Exercise 1.**

**1.** $\frac{2}{3} - \frac{4}{9} = \frac{2}{9}$     **2.** $.42 \times 93 = 39.06$

**3.** $\frac{3}{4} = \frac{9}{12}$     **4.** $720 \text{ km} \div 8 = 90 \text{ km}$

**5.** $\frac{1}{2} + \frac{1}{4} = .75$     **6.** $97\frac{5}{12} - 8\frac{7}{8} = 88\frac{13}{24}$

**7.** $6 = .5 \times 12$     **8.** $\$29.95 - \$5.99 = \$23.96$

**9.** $127 \times 52 = 6604$     **10.** $132 + 19 + 17 = 158$

**11.** $15,640 \div 92 = 170$

**12.** $2 \text{ m} + 2 \text{ m} + 2 \text{ m} + 2 \text{ m} = 8 \text{ m}$

**13.** $3 \text{ hr} + 20 \text{ min} + 4 \text{ hr} + 30 \text{ min} = 7 \text{ hr} + 50 \text{ min}$

**14.** $\$2400 \times 12 = \$28,800$

**Page 8, Exercise 2.**

**1.** 533.58: $502.62 + 27.1 + 3.86 = 533.58$

**2.** 2.8: $14.56 \div 5.2 = 2.8$

**3.** $\frac{7}{12}$: $\frac{3}{4} - \frac{1}{6} = \frac{7}{12}$

**4.** 358: $1520 - 1162 = 358$

**5.** 15: $\frac{1}{5} \times 75 = 15$

**6.** 70%: $7 \div 10 = .7$ or 70%

**7.** $15.66 savings: $\$39.95 - \$24.29 = \$15.66$

**8.** $450 interest: $\$3000 \times .15 = \$450$

**9.** 1 mile: $\frac{2}{3} \times 1\frac{1}{2} = 1$

**10.** $19 each: $\$57 \div 3 = \$19$

**11.** 1524: $127 \times 12 = 1524$

**12.** $\frac{7}{26}$: $\frac{7}{16} \div 1\frac{5}{8} = \frac{7}{26}$

**13.** 540 minutes: $9 \text{ hr} \times 60 \text{ min} = 540 \text{ min}$

**14.** 55: $97 - 42 = 55$

**15.** $8.40: $\frac{\$9.50 + \$7.30}{2} = \$8.40$

**Page 10, Exercise 3.**

**1.** $84 = 84$     **2.** $30 = 30$

**3.** $\frac{5}{6} \neq \frac{6}{5}$     **4.** $.48 = .48$

**5.** $62.962 = 62.962$     **6.** $17 \neq -17$

**7.** $\frac{3}{4} = \frac{3}{4}$     **8.** $20,209 = 20,209$

**9.** $21\frac{7}{16} = 21\frac{7}{16}$     **10.** $\$26.23 \neq -\$26.23$

**Page 11, Exercise 4.**

**1.** $\frac{7}{8} = \frac{7}{8}$     **2.** $7 \neq 15$

**3.** $.01664 = .01664$     **4.** $100 = 100$

**5.** $138\frac{3}{8} = 138\frac{3}{8}$     **6.** $18 \neq 2$

**7.** $5.5 = 5.5 / 5\frac{1}{2} = 5\frac{1}{2}$     **8.** $\$5712 = \$5712$

**9.** $17.28 = 17.28$     **10.** $16.76 = 16.76$

**Page 12, Exercise 5.**

**A. 2.** $5(4-2) = (5 \cdot 4) - (5 \cdot 2)$

$$5(2) = 20 - 10$$
$$10 = 10$$

**3.** $3(9+4) = (3 \cdot 9) + (3 \cdot 4)$

$$3(13) = 27 + 12$$
$$39 = 39$$

**4.** $2(6-3) = (2 \cdot 6) - (2 \cdot 3)$

$$2(3) = 12 - 6$$
$$6 = 6$$

**5.** $6(6+1) = (6 \cdot 6) + (6 \cdot 1)$

$$6(7) = 36 + 6$$
$$42 = 42$$

**6.** $3(20-15) = (3 \cdot 20) - (3 \cdot 15)$

$$3(5) = 60 - 45$$
$$15 = 15$$

**7.** $11(9-5) = (11 \cdot 9) - (11 \cdot 5)$

$$11(4) = 99 - 55$$
$$44 = 44$$

**8.** $12(7+6) = (12 \cdot 7) + (12 \cdot 6)$

$$12(13) = 84 + 72$$
$$156 = 156$$

**9.** $10(3+6) = (10 \cdot 3) + (10 \cdot 6)$

$$10(9) = 30 + 60$$
$$90 = 90$$

**Page 12, Exercise 5.**

**B. 2.** $6(10 + 2) = (6 \cdot 10) + (6 \cdot 2)$
$$6(12) = 60 + 12$$
$$72 = 72$$

**3.** $4(25 + 10) = (4 \cdot 25) + (4 \cdot 10)$
$$4(35) = 100 + 40$$
$$140 = 140$$

**4.** $3(22 + 11) = (3 \cdot 22) + (3 \cdot 11)$
$$3(33) = 66 + 33$$
$$99 = 99$$

**5.** $7(14 + 3) = (7 \cdot 14) + (7 \cdot 3)$
$$7(17) = 98 + 21$$
$$119 = 119$$

**6.** $9(10 - 6) = (9 \cdot 10) - (9 \cdot 6)$
$$9(4) = 90 - 54$$
$$36 = 36$$

**7.** $5(14 + 6) = (5 \cdot 14) + (5 \cdot 6)$
$$5(20) = 70 + 30$$
$$100 = 100$$

**8.** $25(4 - 3) = (25 \cdot 4) - (25 \cdot 3)$
$$25(1) = 100 - 75$$
$$25 = 25$$

**9.** $20(14 + 15) = (20 \cdot 14) + (20 \cdot 15)$
$$20(29) = 280 + 300$$
$$580 = 580$$

**10.** $.5(22 - 19.1) = (.5 \cdot 22) - (.5 \cdot 19.1)$
$$.5(2.9) = 11 - 9.55$$
$$1.45 = 1.45$$

**Page 12, Exercise 6.**
  1. Commutative Property
  2. Distributive Property
  3. Associative Property
  4. Associative Property
  5. Associative Property
  6. Distributive Property
  7. Commutative Property
  8. Commutative Property
  9. Commutative Property
  10. Distributive Property
  11. Associative Property
  12. Distributive Property
  13. Commutative Property
  14. Distributive Property

**Page 13, Exercise 7.**
**A.** 14 because you multiply before adding.

**B. 1.** 14    **2.** 18    **3.** 18
  **4.** 36    **5.** 11    **6.** 2
  **7.** 82    **8.** 7    **9.** 6
  **10.** 10    **11.** 11    **12.** 175
  **13.** 9    **14.** 29    **15.** 7
  **16.** 3    **17.** 5    **18.** 10

# Chapter 2: Variables

**Pages 16-17, Exercise 1.**
  Possibilities, answers may vary.

**A. 1.** $a + b$    **2.** $3r - 2r$    **3.** $\frac{y}{2} \cdot 3$
  **4.** $(2a + b)$    **5.** $3(2b + 3)$    **6.** $3b \cdot r$
  **7.** $6b \div 2b$    **8.** $a(2 + 3)$    **9.** $y - 16$
  **10.** $25 \div c$    **11.** $3 \cdot m + 2$    **12.** $24m + 17m$
  **13.** $8 + d$    **14.** $3b(6 \cdot 4)$    **15.** $c(a + b)$
  **16.** $b - 25$    **17.** $365g$    **18.** $\frac{12}{v}$
  **19.** $12 + x$    **20.** $\frac{x}{x}$

**B. 1.** $6, y, 10$    **2.** $n, 16$
  **3.** $4, a$    **4.** $12, b$
  **5.** $12a, 6b, 4$    **6.** $4, 2, 10$
  **7.** $\frac{1}{2}ab, 3$    **8.** $c, 32$
  **9.** $4x, 2y$    **10.** $2, 3, x$

**C. 1.** c    **2.** d    **3.** a
  **4.** b    **5.** m    **6.** i
  **7.** f    **8.** k    **9.** l
  **10.** e    **11.** o    **12.** n
  **13.** g    **14.** j    **15.** h

**Page 17, Exercise 1 (cont.)**
Wording of answers may vary.
**D. 1.** Twice a number plus 4.
   **2.** Fifty minus some number.
   **3.** Three times the sum of a number plus 3.
   **4.** The sum of three numbers.
   **5.** The difference between 14 and 5 divided by some number.
   **6.** Twenty-four times a number.
   **7.** Nine-eighths of a number minus the sum of 2 plus 4.
   **8.** Eight times a number plus nine times a number.
   **9.** Three times the sum of nine times a number plus 8 times a number.
  **10.** Fourteen minus twelve plus the sum of 3 and a number.

**Page 18, Exercise 2.**

| | | |
|---|---|---|
| **1.** 21 | **2.** 6 | **3.** 189 |
| **4.** 34 | **5.** 21 | **6.** 9 |
| **7.** 3 | **8.** 4 | **9.** 3 |
| **10.** 7 | **11.** 10 | **12.** 46 |
| **13.** 1 | **14.** 2 | **15.** 24 |
| **16.** 14 | **17.** 30 | **18.** 34 |

**Page 20, Exercise 3.**

**1.** $16a$     **2.** $3y$     **3.** $60b + 5$
**4.** $13stu$     **5.** $5a + 5b$     **6.** $15c + 4r$
**7.** $7e + 5$     **8.** $5y + 38$     **9.** $8ab + 26$
**10.** $30k - 13w + 7$     **11.** $-7x^3 + 4x^2 + 3x + 4$
**12.** $3a^2b + ab^2$     **13.** $11x^2 + 5xy$
**14.** $2x^2 + y + 15$     **15.** $2x^2$
**16.** $32c^3d - 9$     **17.** $15x^2y + xy + 3$
**18.** $2mn^2 + m^2n$     **19.** $33k + 33m + 32$
**20.** $-3w + 16x + 24xy$

# Chapter 3: Solving Linear Equations

**Pages 23-24, Exercise 1.**
**A. 1.** $a = 22$     **2.** $z = 35$     **3.** $x = 52$
   **4.** $w = 4.9$     **5.** $g = 7\frac{3}{4}$     **6.** $b = 26.5$
   **7.** $r = 283$     **8.** $v = 3.351$     **9.** $h = 607$
  **10.** $c = 13\frac{7}{8}$

**B. 1.** $k = 73$     **2.** $r = 21$     **3.** $f = 11.97$
   **4.** $x = 20\frac{1}{2}$     **5.** $t = 11.62$     **6.** $p = 1\frac{1}{5}$
   **7.** $y = 15$     **8.** $n = 10\frac{5}{6}$     **9.** $d = 8.1$
  **10.** $s = 611$

**C. 1.** $x = 20$     **2.** $b = 49$     **3.** $f = 60$
   **4.** $a = 72.9$     **5.** $m = 55$     **6.** $d = 625$
   **7.** $v = 42.30$     **8.** $c = 144$     **9.** $n = 150$
  **10.** $n = 230$

**D. 1.** $r = 13$     **2.** $c = 6.2$     **3.** $g = \$.23$
   **4.** $w = 12.34$     **5.** $d = .01$     **6.** $q = 49$
   **7.** $h = \frac{1}{10}$     **8.** $j = .49$     **9.** $x = 21$
  **10.** $w = 12$

**E. 1.** $m = \$34.80$     **2.** $f = 45$     **3.** $b = \frac{1}{15}$
   **4.** $k = 1.86$     **5.** $m = \$496$     **6.** $y = 1$
   **7.** $c = 2.9$     **8.** $t = 24.616$     **9.** $b = \$7.29$
  **10.** $n = 430$

**Pages 26-28, Exercise 2.**
**A. 1.** $n - 5$     **2.** $n + 123$     **3.** $n + 12$
   **4.** $27n$     **5.** $n - .58$     **6.** $4n$
   **7.** $\frac{1}{100}n$     **8.** $15n$     **9.** $n - 10$
  **10.** $10n$     **11.** $\frac{14}{n}$     **12.** $10 - y$
  **13.** $42 + x$     **14.** $\frac{97}{2}$     **15.** $12m$
  **16.** $132 + n$

**B. 1.** $c + 8 = 35$     **2.** $2c = 42$     **3.** $c - 18 = 172$
   **4.** $\frac{1}{4}c = 20$     **5.** $c + 100 = 427$    **6.** $3c = 45$
   **7.** $c - 32 = 47$     **8.** $\frac{c}{20} = 7.5$

**C. 1.** $c = 27$     **2.** $c = 21$     **3.** $c = 190$
   **4.** $c = 80$     **5.** $c = 327$     **6.** $c = 15$
   **7.** $c = 79$     **8.** $c = 150$

**D. 1.** $16b = \$56.96$      $b = \$3.56$
   **2.** $3c = \$34.05$      $c = \$11.35$
   **3.** $\$27.50 + a = \$35.20$      $a = \$7.70$
   **4.** $19.4 + n = 98.5$      $n = 79.1$
   **5.** $m - \$114 = \$428$      $m = \$542$
   **6.** $b - \$1.29 = \$.71$      $b = \$2.00$
   **7.** $\frac{1}{2}t = 36$      $t = 72$
   **8.** $3a = \$37.50$      $a = \$12.50$
   **9.** $6.3 + m = 12$      $m = 5.7$
  **10.** $\frac{1}{3}i = \$528$      $i = \$1584$
  **11.** $74.8 + s = 187$      $s = 112.2$
  **12.** $3m = 39$      $m = 13$
  **13.** $b - 1366 = 1334$      $b = 2700$
  **14.** $j + 5 = 125$      $j = 120$
  **15.** $\frac{1}{3}t = 136$      $t = 408$

**Pages 29-30, Exercise 3.**

**A.** **1.** $x = 2$    **2.** $x = \frac{2}{3}$    **3.** $x = 52$
**4.** $x = 12$    **5.** $x = 48$    **6.** $x = 7$
**7.** $x = 5$    **8.** $x = 8$    **9.** $x = 134.4$
**10.** $x = 8$    **11.** $x = 13$    **12.** $x = 8$
**13.** $x = 6$    **14.** $x = \frac{8}{9}$    **15.** $x = 20$

**B.** **1.** $x = 3$    **2.** $x = 7$    **3.** $x = 15$
**4.** $x = 5$    **5.** $x = 4.1$    **6.** $x = 4$
**7.** $x = 8$    **8.** $x = 10$    **9.** $x = 4$
**10.** $x = \frac{15}{16}$    **11.** $x = 9$    **12.** $x = 2$

**C.** **1.**  $4(2x + 7) = 108$    **2.**  $5(3x + 8) = 175$
$\qquad 8x + 28 = 108$    $\qquad 15x + 40 = 175$
$\qquad 8x = 80$    $\qquad 15x = 135$
$\qquad x = 10$    $\qquad x = 9$

**3.**  $(3x + 2) + (9x - 1) = 37$    **4.**  $2(3x + 14) - 28 = 27$
$\qquad 12x + 1 = 37$    $\qquad 6x + 28 - 28 = 27$
$\qquad 12x = 36$    $\qquad 6x = 27$
$\qquad x = 3$    $\qquad x = 4.5$

**5.**  $3x + 2(5x - 3) = 7$    **6.**  $6(2x - 2) + 8 = 32$
$\qquad 3x + 10x - 6 = 7$    $\qquad 12x - 12 + 8 = 32$
$\qquad 13x = 13$    $\qquad 12x = 36$
$\qquad x = 1$    $\qquad x = 3$

**7.**  $3x + 4(3x - 5) = 25$ **8.**  $2(2x + 7) + (3x - 5) = 37$
$\qquad 3x + 12x - 20 = 25$    $\qquad 4x + 14 + 3x - 5 = 37$
$\qquad 15x = 45$    $\qquad 7x = 28$
$\qquad x = 3$    $\qquad x = 4$

**9.**  $13 + 8(5x - 9) = 21$    **10.**  $(12x + 30) - 2(4x) = 90$
$\qquad 13 + 40x - 72 = 21$    $\qquad 12x + 30 - 8x = 90$
$\qquad 40x = 80$    $\qquad 4x = 60$
$\qquad x = 2$    $\qquad x = 15$

**11.** $(2x - 7) + (7x + 10) = 84$ **12.** $6(5x - 9) + 24 = 120$
$\qquad 9x + 3 = 84$    $\qquad 30x - 54 + 24 = 120$
$\qquad 9x = 81$    $\qquad 30x = 150$
$\qquad x = 9$    $\qquad x = 5$

**Page 32, Exercise 4.**

**1.** $g = 3$    **2.** $x = 4$    **3.** $v = 10$
**4.** $y = 5$    **5.** $z = 9$    **6.** $w = 4$
**7.** $s = 1$    **8.** $m = 1$    **9.** $c = 4$
**10.** $k = 2$    **11.** $x = \frac{1}{3}$    **12.** $d = 5$
**13.** $f = 6$    **14.** $e = 4$    **15.** $n = 3$
**16.** $x = 0$    **17.** $a = 3$    **18.** $r = 11$
**19.** $k = 2$    **20.** $w = 3$    **21.** $a = 10$
**22.** $x = 7\frac{1}{3}$    **23.** $c = 12$    **24.** $p = 30$
**25.** $n = 18$    **26.** $c = 1$    **27.** $p = 31$
**28.** $b = 2$    **29.** $x = 3$    **30.** $y = 7$

## Chapter 4: Word Problems

**Page 33, Exercise 1.**

**1.** $n + 4 = 15$    $n = 11$
**2.** $n - 45 = 82$    $n = 127$
**3.** $7n = 77$    $n = 11$
**4.** $\frac{1}{2}n = 14\frac{1}{2}$    $n = 29$
**5.** $\frac{2n}{6} = 42$    $n = 126$
**6.** $\frac{45}{n} = 5$    $n = 9$
**7.** $\frac{n+12}{3} = 9$    $n = 15$
**8.** $4(n - 2) = 14$    $n = 5.5$
**9.** $4 + 2n = 52$    $n = 24$
**10.** $5n - 12 = 9$    $n = 4.2$

**Pages 35-36, Exercise 2.**
Chosen variables will differ.

**1.** $c = \frac{\$211}{4}$    $c = \$52.75$
**2.** $c = 15{,}000 - 792$    $c = 14{,}208$
**3.** $n = 2(47) - 12$    $n = 82$
**4.** $f = \$15.20 - \$2.59 - \$4.30$    $f = \$8.31$
**5.** $d = \frac{\$3513.02}{23}$    $d = \$152.74$
**6.** $d = 2009 - 1986$    $d = 23$
**7.** $b = 72" - 6"$    $b = 66"$
**8.** $g = 3(\$21.50 + \$14.50)$    $g = \$108$
**9.** $s = \frac{\$695}{4}$    $s = \$173.75$
**10.** $w = \$62(7)$    $w = \$434$
**11.** $p = \$3.90\frac{4}{3}$  or  $p = \frac{\$3.90}{\frac{3}{4}}$    $p = \$5.20$
**12.** $c = \frac{\$12{,}500 - 500}{15}$    $c = \$800$
**13.** $g = \frac{66}{6}$    $g = 11$
**14.** $a = 1350 - 360 - 720$    $a = 270$
**15.** $m = \frac{200\text{km}}{1.6\text{km}}$    $m = 125$
**16.** $t = 42 + 37$    $t = 79$
**17.** $s = 8(\$1.25) - 4(\$2.25)$    $s = \$1$
**18.** $s = \$64.35 + \$12.95 + \$147.20 + \$24.95$
$\qquad s = \$249.45$
$\qquad b = \$892.57 - \$249.45$    $b = \$643.12$
**19. a.**  $l = \frac{30}{4}$    $l = 7.5$
**b.**  $f = 30 - 7.5$    $f = 22.5$
    c.  $t = \left(30 \cdot \frac{2}{3}\right) - 7.5$  $f = 12.5$

**Pages 38-39, Exercise 3.**

**1.** $n + (n + 26) = 80$
   $n = 27$
   $n + 26 = 53$
**3.** $n + (n - 12) = 30$
   $n = 21$
   $n - 12 = 9$
**5.** $n + (2n - 1) = 41$
   $n = 14$
   $2n - 1 = 27$
**7.** $n + 4n = 65$
   $n = 13$
   $4n = 52$
**9.** $n + (n + 7) = 33$
   $n = 13$ men
   $n + 7 = 20$ women
**11.** $n + (n + 15) = 41$
   $n = 13$
   $n + 15 = 28$
**13.** $42n + 1.5n(7) = \$315$
   $n = \$6$
   $1.5n = \$9$
**15.** $n + 5n = 18$
   $n = 3$
   $5n = 15$

**2.** $n + (n + 26) = 80$
   $n = \$27$
   $n + 26 = \$53$
**4.** $n + (n + 12) = 38$
   $n = 13$
   $n + 12 = 25$
**6.** $n + 4n = 90$
   $n = 18$
   $4n = 72$
**8.** $n + (n - 2.5) = 121.5$
   $n = 62$ kg
   $n - 2.5 = 59.5$ kg
**10.** $n + (7n + 8) = 32$
   $n = 3$
   $7n + 8 = 29$
**12.** $n + (n - 1) = 5$
   $n = 3$ m
   $n - 1 = 2$ m
**14.** $n + (n + 13) = 91$
   $n = 39$
   $n + 13 = 52$

**Pages 42-44, Exercise 4.**

**A. 1.** $n + (n + 1) = 285$
   $n = 142$
   $n + 1 = 143$
**2.** $n + (n + 2) = 86$
   $n = 42$
   $n + 2 = 44$
**3.** $n + (n + 1) + (n + 2) = 138$
   $n = 45$
   $n + 1 = 46$
   $n + 2 = 47$
**4.** $n + (n + 2) + (n + 4) = 111$
   $n = 35$
   $n + 2 = 37$
   $n + 4 = 39$
**5.** $n + (n + 1) + (n + 2) + (n + 3) = 262$
   $n = 64$
   $n + 1 = 65$
   $n + 2 = 66$
   $n + 3 = 67$

**B. 1.** $d + (12 - d) = 12$ (coins)
   $.10d + .05(12 - d) = \$.85$
   $d = 5$ (dimes)      $5 \cdot \$.10 = \$.50$
   $12 - d = 7$ (nickels)  $7 \cdot \$.05 = \$.35$
**2.** $n + n + n = 3n$ (coins)
   $.05n + .10n + .25n = \$2.00$
   $n = 5$ (nickels)   $5 \cdot \$.05 = \$.25$
   $n = 5$ (dimes)    $5 \cdot \$.10 = \$.50$
   $n = 5$ (quarters)  $5 \cdot \$.25 = \$1.25$

**3.** $3n + n = 4n$ (coins)
   $.05(3n) + .10n = \$8.50$
   $n = 34$ (dimes)     $34 \cdot \$.10 = \$3.40$
   $3n = 102$ (nickels)  $102 \cdot \$.05 = \$5.10$
**4.** $n + 5n = 6n$ (coins)
   $.10n + .25(5n) = \$16.20$
   $n = 12$ (dimes)     $12 \cdot \$.10 = \$1.20$
   $5n = 60$ (quarters) $60 \cdot \$.25 = \$15.00$
**5.** $n + (n + 2) + 3n = 5n + 2$
   $.10n + .05(n + 2) + .01(3n) = \$.64$
   $n = 3$ (dimes)      $3 \cdot \$.10 = \$.30$
   $n + 2 = 5$ (nickels)  $5 \cdot \$.05 = \$.25$
   $3n = 9$ (pennies)    $9 \cdot \$.01 = \$.09$

**C. 1.** $x =$ width in meters
   $3x - 4 =$ length in meters
   $2(x) + 2(3x - 4) = 224$
   $x = 29$ meters (width)
   $3x - 4 = 83$ meters (length)
**2.** Perimeter of square = perimeter of triangle
   24 meters = 24 meters
   $P = 4s$
   24 meters = $4s$
   6 meters = $s$
   1 side = 6 meters
**3.** $l =$ length in centimeters
   $w = \frac{1}{2}l - 3$ in centimeters
   $2(l) + 2(\frac{1}{2}l - 3) = 60$
   $l = 22$ centimeters (length)
   $w = \frac{1}{2}l - 3 = 8$ centimeters (width)
**4.** Side 1 = $2t$
   Side 2 = $t$
   Side 3 = $3t - 20$
   $2t + t + (3t - 20) = 106$
   $t = 2$ m (side 2)
   $2t = 42$ m (side 1)
   $3t - 20 = 43$ m (side 3)
**5.** $\angle + \angle + \angle = 180°$
   $\angle 1 = a$
   $\angle 2 = a - 20°$
   $\angle 3 = 2(a - 20°)$
   $a + (a - 20°) + 2(a - 20°) = 180°$
   $a = 60°$
   $\angle 1 = 60°$
   $\angle 2 = 40°$
   $\angle 3 = 80°$

**D. 1.** $n + (n + 2) + (n + 2 + 5) = 33$ meters
   $n = 8$ meters (first side)
   $n + 2 = 10$ meters (second side)
   $n + 2 + 5 = 15$ meters (third side)
**2.** $n + (2n + 5) + (3n - 1) = \$12.25$
   $.05n + .10(2n + 5) + .25(3n - 1) = \$12.25$
   $n = 12$ (nickels)     $12 \cdot \$.05 = \$.60$
   $2n + 5 = 29$ (dimes) $29 \cdot \$.10 = \$2.90$
   $3n - 1 = 35$ (qtrs.)  $35 \cdot \$.25 = \$8.75$

3. $n + (n + 1) + (n + 2) = 6n$
   $n = 1$
   $n + 1 = 2$
   $n + 2 = 3$
4. Step 1:                  Step 2:
   $w$ = width              $A = lw$
   $3w$ = length            $l = 9$
   $2(w) + 2(3w) = 24$      $w = 3$
   $w = 3$                  $A = 27$ feet
   $3w = 9$
5. $m + (m + 5) + (m + 10) = 120$
   $m = 35$
   $m + 5 = 40$
   $m + 10 = 45$
6. $n + 2n + 2(2n) = 70$  (coins)
   $n = 10$ (nickels)        $10 \cdot \$.05 = \$.50$
   $2n = 20$ (dimes)         $20 \cdot \$.10 = \$2.00$
   $2(2n) = 40$ (qtrs.)      $40 \cdot \$.25 = \$10.00$
                            Total = $12.50
7. $\angle + \angle + \angle = 180°$
   $n + n + 2n = 180°$
   $n = 45°$ (first angle)
   $n = 45°$ (second angle)
   $2n = 90°$ (third angle)
8. $n + (n + 1) = (n + 2) + 35$
   $n = 36$
   $n + 1 = 37$
   $n + 2 = 38$
9. $.50n + .19(n + 20) + .29 \cdot 2(n + 20) = \$30.64$
   $n = 12$                  $12 \cdot \$.50 = \$6.00$
   $n + 20 = 32$            $32 \cdot \$.19 = \$6.08$
   $2(n + 20) = 64$         $64 \cdot \$.29 = \$18.56$

## Chapter 5: Formulas

### Pages 46-47, Exercise 1.

A. 1. $w = \frac{A}{l}$           2. $P = \frac{I}{RT}$

   3. $r = \frac{D}{t}$           4. $l = \frac{V}{wh}$

   5. $b = \frac{2A}{h}$          6. $C = \frac{5}{9}(F - 32)$

   7. $h = \frac{V}{\pi r^2}$     8. $r = \frac{C}{2\pi}$

   9. $l = \frac{P-2w}{2}$        10. $d = \frac{C}{\pi}$

   11. $s = \frac{P}{4}$          12. $b = \frac{3V}{h}$

   13. $a^2 = c^2 - b^2$          14. $h = \frac{2A}{b}$

   15. $l = \frac{A}{w}$

B. 1. $b = 16$ units             2. $t = 2$ hours 15 minutes
   3. $h = 27$ feet             4. $w = 15.6$ square yards
   5. $t = 2$ years             6. $h = 6$ yards, yes
   7. $l = 52.5$ feet           8. $d = 25$ feet
   9. $w = 27$ meters           10. $h = 17.4$ meters
   11. $w = 132$ feet           12. $h = 1.75$ inches

### Pages 48-49, Exercise 2.
Variables and formula may differ.
1. $m$ = miles driven = ?
   $x$ = miles per gallon = 23.2
   $g$ = gallons used = 32.5
   $m = xg$
   $m = 754$
2. $c$ = cost per mile driven = ?
   $m$ = miles driven = 300
   $g$ = gallons used = 15
   $p$ = cost per gallon = $1.20
   $c = \frac{gp}{m}$
   $c = \$.06$
3. $m$ = miles driven = ?
   $p$ = cost per gallon = $1.25
   $c$ = cost per mile = $.08
   $g$ = gallons used = 10
   $m = \frac{gp}{c}$
   $m = 156.25$
4. $s$ = money saved = ?
   $c$ = original cost = $129.95
   $d$ = discount = 20%
   $s = cd$
   $s = \$25.99$
5. $d$ = discount received = ?
   $c$ = original cost = $25.50
   $s$ = savings = $4.08
   $d = \frac{s}{c}$
   $d = 16\%$
6. $g$ = cost per gram = ?
   $c$ = cereal cost = $6.44
   $w$ = weight = 920 grams
   $g = \frac{c}{w}$
   $g = \$.007$ gram
7. $c$ = cost of cereal = ?
   $w$ = weight = 920 grams
   $g$ = cost per gram = $.006
   $c = wg$
   $c = \$5.52$
8. $l$ = length of fifth side = ?
   $a$ = side 1 = 27.5 meters
   $b$ = side 2 = 132 meters
   $c$ = side 3 = 29.5 meters
   $d$ = side 4 = 133 meters
   $p$ = perimeter = 398 meters
   $l = p - (a + b + c + d)$
   $l = 76$ meters

9. $s$ = monthly salary = ?
   $a$ = average calls per week = 527
   $p$ = paid per call = $.17
   $w$ = weeks per month = 4
   $s = apw$
   $s$ = $358.36

10. $a$ = what average worker earns = ?
    $e$ = employees = 12
    $p$ = monthly payroll = $17,131.44
    $a = \frac{p}{e}$
    $a$ = $1427.62

11. $y$ = yearly salary = ?
    $m$ = monthly salary = $1427.62
    $t$ = months per year = 12
    $y = mt$
    $y$ = $17,131.44

12. $a$ = amount put aside = ?
    $y$ = yearly salary = $17,131.44
    $r$ = retirement % of salary = 12%
    $e$ = employees = 12
    $a = yre$
    $a$ = $24,669.24

13. $a$ = cost of 2 meals = ?
    $x$ = cost of one meal = $12.50
    $y = \frac{1}{2}x$ = $6.25
    $a = x + y$
    $a$ = $18.75

14. $c$ = cost of fence = ?
    $l$ = perimeter length = 35 meters
    $w$ = perimeter width = 28 meters
    $p$ = price per meter = $51
    $c = p(2l + 2w)$
    $c$ = $6426

15. $c$ = commission = ?
    $s$ = monthly sales = $24,572.27
    $r$ = commission rate = 12%
    $c = sr$
    $c$ = $2948.67

16. Step 1
    $t$ = gain for elapsed time = ?
    $e$ = elapsed time = 9.5 hours
    $g$ = gain per hour of elasped time = 6 minutes
    $t = eg$
    $t$ = 57 minutes

    Step 2
    $a$ = exact time = ?
    $w$ = when Phyllis looked at watch = 6:30 PM
    $t$ = gain for elapsed time = 57 minutes
    $a = w - t$
    $a$ = 5:33 PM

17. Step 1
    $s$ = monthly salary = ?
    $a$ = amount spent = $500
    $\frac{1}{5}s = a$
    $s$ = $2500

Step 2
$i$ = cost of insurance = ?
$p$ = % of monthly salary = 60%
$s$ = monthly salary = $2500
$i = ps$
$i$ = $1500

18. $t$ = tax savings = ?
    $p$ = present tax rate = 30%
    $n$ = new tax rate = 10%
    $a$ = annual income = $40,000
    $t = p(a) - n(a)$
    $t$ = $8000

19. $m$ = minimum number of games = ?
    $s$ = season ticket = $255
    $i$ = individual tickets = $15
    $s = \frac{s}{i}$
    $m$ = 17

20. Step 1
    $g$ = cost of gloves for entire family = ?
    $m$ = number of family members = 12
    $p$ = price per pair of gloves = $15
    $g = mp$
    $g$ = $180

    Step 2
    $t$ = total of gloves and tax = ?
    $g$ = cost of gloves for entire family = $180
    $s$ = sales tax rate = 14%
    $t = gs + g$
    $t$ = $205.20